RENEWALS 458-4574
DATE DUE

**WITHDRAWN
UTSA Libraries**

Two Centuries of Compensation for U.S. Production Workers in Manufacturing

PREVIOUS BOOKS OF LAWRENCE H. OFFICER

An Econometric Model of Canada under the Fluctuating Exchange Rate, 1968.

The International Monetary System: Problems and Proposals (co-editor), 1969.

Canadian Economic Problems and Policies (co-editor), 1970.

Supply Relationships in the Canadian Economy: An Industry Comparison, 1972.

Issues in Canadian Economics (co-editor), 1974.

The Monetary Approach to the Balance of Payments: A Survey (co-author), 1978.

So You Have to Write an Economics Term Paper... (co-author), 1981.

Purchasing Power Parity: Theory, Evidence and Relevance, 1982.

International Economics (editor), 1987.

Between the Dollar-Sterling Gold Points: Exchange Rates, Parities and Market Behavior, 1791–1931, 1996; paperback reissue, 2007.

Monetary Standards and Exchange Rates (co-editor), 1997.

Pricing Theory, Financing of International Organisations and Monetary History, 2007.

Two Centuries of Compensation for U.S. Production Workers in Manufacturing

Lawrence H. Officer

TWO CENTURIES OF COMPENSATION FOR U.S. PRODUCTION WORKERS IN MANUFACTURING
Copyright © Lawrence H. Officer, 2009.

All rights reserved.

First published in 2009 by
PALGRAVE MACMILLAN®
in the United States—a division of St. Martin's Press LLC,
175 Fifth Avenue, New York, NY 10010.

Where this book is distributed in the UK, Europe and the rest of the world, this is by Palgrave Macmillan, a division of Macmillan Publishers Limited, registered in England, company number 785998, of Houndmills, Basingstoke, Hampshire RG21 6XS.

Palgrave Macmillan is the global academic imprint of the above companies and has companies and representatives throughout the world.

Palgrave® and Macmillan® are registered trademarks in the United States, the United Kingdom, Europe and other countries.

ISBN-13: 978–0–230–61566–3
ISBN-10: 0–230–61566–X

Library of Congress Cataloging-in-Publication Data

Officer, Lawrence H.
  Two centuries of compensation for U.S. production workers in manufacturing / Lawrence H. Officer.
    p. cm.
  Includes bibliographical references and index.
  ISBN 0–230–61566–X
    1. Wages—Manufacturing industries—United States. I. Title.

HD4975.O44 2009
331.2'10973—dc22   2008040480

A catalogue record of the book is available from the British Library.

Design by Newgen Imaging Systems (P) Ltd., Chennai, India.

First Edition: May 2009

10 9 8 7 6 5 4 3 2 1

Printed in the United States of America.

To
Sandra Diane Officer

Truth needs no color—Beauty no pencil.
                                        Shakespeare

Contents

List of Figures — viii
List of Tables — ix
Preface — xi
List of Abbreviations and Symbols — xiii

1. Methodology — 1
2. Data Sources — 15
3. Existing Earnings and Wage Series — 75
4. Plan for Construction of New Series — 97
5. Average Hourly Earnings — 103
6. Average Hourly Benefits — 155
7. Nominal Compensation, Real Compensation, and Standard of Living — 165

Appendix: Feeder Series — 181
References — 195
Index of Names — 209
Index of Subjects — 213

Figures

7.1	Average hourly compensation	168
7.2	Ratio of benefits to compensation	169
7.3	Real average hourly compensation	171
7.4	Work-hours: Actual and required-to-purchase-consumer-bundle	174
7.5	Real average hourly compensation: Cycle component	179

Tables

2.1	Average hourly earnings or hourly wage rates, production workers in manufacturing: BLS Series	16
2.2	BLS hourly earnings or wage series with fixed weights, production workers in manufacturing	24
2.3	Ratio of hours at work to hours paid, 1947–2005, production and nonsupervisory employees, manufacturing: BLS Data	58
2.4	Total benefits in manufacturing—Data sources	70
3.1	Average hourly earnings or hourly wage rate, production workers in manufacturing: Private uniform series, 1890–1935	76
3.2	Average daily wage rate, production workers in manufacturing: existing series, 1860–1890	82
3.3	Average hourly earnings or hourly wage rate, production workers in manufacturing: Composite series	88
3.4	Average annual earnings, interpolation of intercensal years, wage-earners in manufacturing	92
5.1	Average hours per week, 1920–1932: Production workers in manufacturing	106
5.2	Average annual earnings, existing studies, adjusted Census data, 1859–1919	110
5.3	Hand and custom trades eliminated from 1849–1889 Census figures	114
5.4	Hand and custom trades retained in 1849–1889 Census figures	115
5.5	Computation of average annual earnings, manufacturing: 1849 and 1859	115
5.6	Computation of average annual earnings, manufacturing: 1869, 1879, and 1889	117
5.7	Average wage rate in antebellum period—Northeast, wage-earners in manufacturing	120

5.8	Average number of days of operation—Manufacturing, Census years, 1849–1919	122
5.9	Computation of average annual earnings, manufacturing, Northeast, 1849—Census data	130
5.10	Rees interpolator series and Douglas payroll series: Comparison of coverage	134
5.11	Ratios to extend coverage of new series	141
5.12	Estimated average wage of females relative to adult males, manufacturing, Northeast, 1815–1859	148
5.13	Interpolation of average hourly earnings between benchmark years	153
6.1	Comparison of benefits mark-up over earnings, manufacturing: Present study versus Bureau of Economic analysis, 1957–1988	157
6.2	Benefits component of average hourly compensation: 1906–1929	162
7.1	Average hourly compensation, earnings, and benefits: 1800–2006	166
7.2	Real average hourly compensation: 1800–2006	170
7.3	Work-hours required to purchase consumer bundle: 1900–2006	173
7.4	Ratio of actual to consumer-bundle-required work-hours: 1904–2006	175
A.1	Correspondence of *Bulletin 18* and Census occupations	182
A.2	Wage-earners, manufacturing, by region, 1820–1859	185
A.3	Computation of adjusted wage-earners, manufacturing, Northeast, 1859	185
A.4	Estimated age-sex distribution of workers in manufacturing, Northeast, 1800–1859	188
A.5	Average hours per day, production workers in manufacturing: Comparison of new series with existing series, 1830–1890	191

PREFACE

This book began largely as a reaction to the Millennial edition of *Historical Statistics of the United States*, edited by Susan B. Carter, Scott Sigmund Gartner, Michael R. Haines, Alan L. Olmstead, Richard Sutch, and Gavin Wright. I noticed that an omission from that impressive five-volume set was a composite long-run series of the average hourly earnings or wages of production workers in manufacturing. That omission seemed strange, because the authors compile long-run composite series for other economic variables, such as GDP and the consumer price index.

Therefore I began to generate the missing series; but I quickly discovered that the task was a giant one. There was good reason for the lack of a long-run earnings or wage series for manufacturing production workers: the series requires considerable effort and enterprise for its construction. This book contains no new archival research; rather, the results of many previous works relating to wages and related variables are assembled and synthesized in a meaningful way. Yet the task took most of my research time of two years to accomplish the objective.

As I digested and integrated the studies of previous authors, my admiration for those who did original archival investigation and reported their results grew substantially. The influence of authors such as Donald R. Adams Jr., Jeremy Atack, Fred Bateman, Philip R. P. Coelho, Joseph H. Davis, Claudia Goldin, Robert A. Margo, James F. Shepherd, Kenneth Sokoloff, and Georgia C. Villaflor is apparent in the very content of this book. Also, I cannot overestimate my intellectual debt to the great Paul H. Douglas, Clarence D. Long, and Albert Rees—the pioneers in the development of earnings and wage series for production workers in manufacturing—as well as to the great Stanley Lebergott.

The book will be of use to specialists in Economic History and Labor Economics. Hopefully, it will also be of interest to government economists and even educated laypersons. The work is exposited in a way to enhance understanding by all these groups. The reader may notice that there are no footnotes or endnotes. That is deliberate. Hopefully, the adopted style will enhance readability.

I thank Robert Whaples for helpful suggestions. The expositions, descriptions, interpretations, opinions, conclusions, judgments, and other unquoted statements in this book are those of the author alone.

LAWRENCE H. OFFICER
Glencoe, Illinois
September 2008

Abbreviations and Symbols

A	artisan wage, for given region
AAE	average annual earnings
AAE_{NE}	average annual earnings, Northeast
AAE_R	average annual earnings, rest-of–United States
$AAE(SV)_{AM,NE}$	average annual earnings, adult males, Northeast, Sokoloff-Villaflor data
ABM	Atack, Bateman, and Margo
ADE	average daily earnings
ADE(SV)	average daily earnings, United States, Sokoloff-Villaflor data
$ADE(SV)_{AM,NE}$	average daily earnings, adult males, Northeast, Sokoloff-Villaflor data
$ADE(SV)_{NE}$	average daily earnings, all workers, Northeast, Sokoloff-Villaflor data
ADH	average daily hours
ADO	average number of days of operation per year
ADW	average daily wage(s)
AHB	average hourly benefits
AHC	average hourly compensation
AHCR	real average hourly compensation
AHE	average hourly earnings
AHE_S	average hourly earnings, unlinked 1800–1919 segment
AHW	average hourly wage(s)
AHWRI	average hourly wage-rate index
ALADW	adjusted Long-Aldrich average daily wage
ALAHW	adjusted Long-Aldrich average hourly wage
AR	adjustment ratio: U.S./Northeast wage ratio
ARB	adjustment ratio: *Bulletin 18* U.S./Northeast wage ratio
ARCS	adjustment ratio: Coelho-Shepherd U.S./Northeast wage ratio
ASM	Annual Survey of Manufactures

AWE	average weekly earnings
AWH	average weekly hours
BEA	Bureau of Economic Analysis
BLS	Bureau of Labor Statistics
CAHCR	cyclical component of real average hourly compensation
CC	U.S. Chamber of Commerce
CES	Current Employment Statistics
CL	Commissioner of Labor
COM	Census of Manufactures
CPHM	Composition of Payroll Hours in Manufacturing
CPI	consumer price index
CS	Coelho and Shepherd
CWD	Current Wage Developments
DAHE	Douglas payroll average hourly earnings
DC	District of Columbia
E_{AM}	adult males as proportion of all employed workers in Northeast manufacturing
$E_{AM,M}$	adult males as proportion of employed male workers in Northeast manufacturing
E_B	boys as proportion of all employed workers in Northeast manufacturing
$E_{B,M}$	boys as proportion of employed male workers in Northeast manufacturing
E_F	females as proportion of all employed workers in Northeast manufacturing
E_M	males as proportion of all employed workers in Northeast manufacturing
E_{NE}	Northeast proportion of U.S. manufacturing employment
E_R	rest-of–United States proportion of U.S. manufacturing employment
ECEC	Employer Costs for Employee Compensation
ECI	employment cost index
EEEC	Employer Expenditures for Employee Compensation
FADW	reconstructed Falkner average daily wage
HACT	actual number of annual work-hours per worker
HEI	hourly earnings index
HP	Hodrick-Prescott
HVCB	number of hours required to purchase consumer bundle
HWS	Hours-at-Work Survey
IPCW	index number of proportion of workers covered by pension plan

IPREM	index number of workers'-compensation premiums
JKO	Jablonski, Kunze, and Otto
log	natural logarithm
M	Margo wage (for given occupation and region)
MV	Margo-Villaflor wage (for given occupation and region)
NAICS	North American Industry Classification System
NCS	National Compensation Survey
NICB	National Industrial Conference Board
NSC	National Safety Council
PCW	proportion of workers covered by pension plans
PM	benefits/earnings ratio
PREM	workers-compensation premiums
PWB	pension-and-welfare benefits
R	ratio of desired series (Y) to interpolator series (X)
RAAE	ratio of rest-of–United States to Northeast average annual earnings, manufacturing, Census data
RAY	ratio of "average number of days of operation per year" (ADO) to "value added in manufacturing, cyclical component" (VC)
RLADH	revised Long-Aldrich average daily hours
RLADW	revised Long-Aldrich average daily wage, Northeast
RMAR	ratio of rest-of–United States to Northeast average daily wage (Margo data)
RSW	Ransom, Sutch, and Williamson
RWUN	U.S./Northeast wage ratio
SA	South Atlantic unskilled wage
SC	South Central unskilled wage
SEEEC	Survey of Employer Expenditures for Employee Compensation
SIC	Standard Industrial Classification System
SV	Sokoloff and Villaflor
TAHCR	trend component of real average hourly compensation
U	unskilled wage, for given region
V	value-added in manufacturing
VC	value-added in manufacturing, cyclical component
VCB	value of consumer bundle
VS	value-added in manufacturing, trend (smoothed) component
W	all-worker wage
W_{AF}	adult-female wage
W_{AM}	adult-male wage

W_B	boy wage
W_F	female wage
W_G	girl wage
W_M	male wage
W_{NE}	Northeast wage
W_R	rest-of–United States wage
W_{RE}	real wage
WA	Adams (male) wage
WAZHMNE	Adams-Zabler average hourly wage, male, Northeast
WAZHNE	Adams-Zabler average hourly wage, all workers, Northeast
WAZHUS	Adams-Zabler average hourly wage, United States
WAZMNE	Adams-Zabler average monthly wage, male, Northeast
WCB	workers'-compensation benefits
WNE	Northeast wage
WNEA	Northeast artisan wage
WNEB	Northeast wage, *Bulletin 18*
WPRA	average of Midwest and South Central artisan wage
WR	rest-of–United States wage
WRA	artisan wage, rest-of–United States
WRB	rest-of–United States wage, *Bulletin 18*
WRU	unskilled wage, rest-of–United States
WSCA	South Central artisan wage
WSCU	South Central unskilled wage
WSNE	skilled wage, Northeast
WSR	skilled wage, rest-of–United States
WUNE	unskilled wage, Northeast
WUR	unskilled wage, rest-of–United States
WUS	U.S. wage
WUSB	U.S. wage, *Bulletin 18*
WZ	Zabler (male) wage
X	interpolator series
Y	desired series (via interpolation)
Z	arbitrary variable

CHAPTER 1

Methodology

The objective of this book is to generate a long-run continuous annual series of *average hourly compensation* for *production workers in manufacturing* in the United States. The series begins in the year 1800, continues to the present, and in principle is ongoing into the future. The study provides full information on construction of the new series; it also puts this construction in context, by reviewing data sources (chapter 2) and the existing literature on related historical series (chapters 3 and 5).

There are several reasons why a book devoted to this series is warranted. First, the generation of such a series has never been performed. That is indeed an unusual situation, as long-run time series for many other important U.S. macroeconomic variables have been developed—all beginning no later than 1900, some even earlier than 1800—with certain of these series continuing to the present and in principle beyond. For example, there are the GDP series assembled by Richard Sutch (2006c); the national-product series compiled by Paul W. Rhode and Richard Sutch (2006); the ongoing GDP series of Louis D. Johnston and Samuel H. Williamson (2008); the Goldsmith saving series compiled by Susan B. Carter, John A. James, and Richard Sutch (2006), and by John A. James and Richard Sylla (2006); the industrial-production index of Joseph H. Davis (2004); the unskilled-wage series of Paul A. David and Peter Solar (1977); the consumer price indexes of Christopher Hanes, Peter H. Lindert, Robert A. Margo, and Richard Sutch (Lindert and Sutch, 2006); and the ongoing consumer price index of Lawrence H. Officer (2007a, 2008a). It behooves one to fill this gap in the literature.

Second, over time, benefits have become an important component of worker compensation. Inclusion of benefits in a long-run compensation series (chapter 6) fills a gap in the literature and enables examination of the evolution of the benefits proportion of compensation (chapter 7).

Third, the pioneering works on earnings and wages of production workers in manufacturing have not been systematically assessed and integrated. These are the studies principally of Paul H. Douglas (1930), Clarence D. Long (1960), and Albert Rees (1960, 1961). Fourth, since the time these authors wrote, data of relevance to developing the desired long-run series have been generated by later scholars, in particular, Donald R. Adams Jr., Jeremy Atack, Fred Bateman, Philip R. P. Coelho, Joseph H. Davis, Claudia Goldin, Robert A. Margo, James F. Shepherd, Kenneth Sokoloff, Georgia C. Villaflor, Samuel H. Williamson, and Jeffrey F. Zabler. The pioneering and later works are all involved in the creation of the new series.

Fifth, production workers (or wage-earners) in manufacturing continue to be an important group in the economy—so important that data regarding these workers have always been reported separately to the U.S. Census and published as such. Census data show that production workers constituted 93 percent of all manufacturing employees in 1899. Even though there has been a downward trend in this figure, the percentage was still as high as 71 percent in 2006. What Albert Rees (1959, p. 12) wrote half a century ago remains largely true: "manufacturing wage earners...[are] one of the largest groups of workers in the economy, one that has been very heavily affected by the growth of unionism in the past twenty-five years, and one of the very few groups for which we have reasonably consistent earnings figures covering a long period."

Sixth, the completed series has implications for our understanding of the standard of living of workers over time. Some original techniques and results of this nature are provided in chapter 7.

Seventh, the presentation of the methodological issues (current chapter) and the description of the procedure of constructing the new series (chapters 4–6) could serve as a template for scholars doing similar work for other countries.

The remainder of this section is devoted to methodological issues, including definition of concepts, identification of characteristics of the new series, and selection from alternative concepts and characteristics.

Definition of Manufacturing

It is necessary to delimit sectoral coverage of the series, so that all "manufacturing," and nothing but manufacturing, is incorporated.

METHODOLOGY ❖ 3

There are two complementary definitions of manufacturing: technological and organizational. One "official" (U.S. Census Bureau) *technological definition* is as follows:

> The Manufacturing sector... comprises establishments engaged in the mechanical, physical, or chemical transformation of materials, substances, or components into new products. (American FactFinder Help, undated-b, p. 1)

The Census Bureau amends the definition to exclude only construction:

> The assembling of component parts of manufactured products is considered manufacturing, except in cases where the activity is appropriately classified in... Construction (American FactFinder Help, undated-b, p. 1)

However, a careful technological definition eliminates all nonmanufacturing industries. As Easterlin (1957, p. 638) writes

> Manufacturing is that range of productive operations entailing a change in form of material goods, and into which natural resources enter only through the provision of a site.... The term "change in form" eliminates transportation and distributive activities, such as wholesaling and retailing. The provision of services, e.g., barbering, medical, and legal services, is excluded by restriction to "material" goods. Finally, agricultural, forestry, fishing, mining, and construction activities are eliminated by the concluding phrase, since in all of these natural resources not only provide a site but enter as an integral part of the production process.

The *organizational definition* eliminates non-factory production, in particular, hand (and/or custom) trades. North (1899, p. 272) presents this point of view with precision:

> Manufacturing must necessarily be treated as comprising the industries carried on under the factory system, which means something entirely different from household industry, from shopwork, from employment at a trade, even when the trade workmen are employed at wages by large contractors.

While in principle the division between hand trades and factory establishments is well defined, the distinction may not be so clear-cut in practice. In fact, following Easterlin (1957, p. 640), there are five criteria that may be used to distinguish hand trades from factories:

hand-work using tools versus production using machinery, no use of power versus use of power, no division of labor versus division of labor, small size versus large size (with size measured by value of output or number of employees), and production for the neighborhood versus production for the general market ("neighborhood establishment" versus "factory establishment").

At one time the U.S. Census Bureau apparently adopted the fifth distinction: "the Census Bureau...settled upon production for the general market as the basis for distinguishing establishments operating under the factory system from the hand trades" (Easterlin, 1957, pp. 640–641). To quote a contemporary Census statement:

> The essential difference between true factories and neighborhood establishments seems to be that the products of factories are distributed beyond the narrow limits of the communities in which they are located, while the products of neighborhood establishments are consumed by local patrons. (Bureau of the Census, 1907, p. xxi)

Interestingly, a recent Census definition appears to belie that statement:

> However, establishments that transform materials or substances into new products by hand or in the worker's home and those engaged in selling to the general public products made on the same premises from which they are sold, such as bakeries, candy stores, and custom tailors, may also be included in this [the manufacturing] sector. (American FactFinder Help, undated-b, p. 1)

A custom trade is a recipient of "the practice of customarily resorting to a particular shop, place of entertainment, etc. to make purchases or give orders" (*Oxford Online Dictionary*). In other words, a custom trade is a neighborhood trade, which may be subsumed under hand trades or considered separately, depending on how broadly one defines hand trades.

Definition of Production Workers

There are three complementary definitions of production versus nonproduction workers.

The first definition is based on the time frequency for which compensation is determined. Traditionally, production workers are synonymous with "*wage-earners*," while nonproduction workers are "salaried employees": production workers receive wages, nonproduction workers salaries. Wages are payment per high-frequency time spent at work, historically per day and then per hour (this category also includes "piecework wages," in which wages are expressed per unit of output rather than per unit of time); salaries are payment scheduled at regular and lower-frequency time intervals, historically per month or week.

The second definition is oriented to the location of work. Nonproduction workers are office workers, because the office is where they are based (although their work may involve locations outside the office); production workers are *nonoffice employees*, they work in the plant or factory. Related to this definition are the terms "white-collar workers" (for nonproduction, office workers) and "*blue-collar workers*" (for production, nonoffice workers). Location determines the nature of clothing worn, and the description of clothing distinguishes the type of worker.

The third definition emanates from the nature of the work. Production workers perform manual work; they are involved with tools, machines, or equipment—either through direct use or via caring for such items. Production *(manual) workers* are distinguished from nonproduction workers, those not involved with tools, machines or equipment. Nonproduction workers include clerks, superintendents, managers, executives, salespeople, and others providing professional or servicing functions. Easterlin (1957, p. 660) makes use of Census documents to conclude that the third definition is primary: "It is clear that the character of work done is the principal criterion for distinguishing wage earners from salaried employees, and not the unit of time which is the basis of compensation." Both the Census Bureau and Bureau of Labor Statistics (BLS) adhere to this definition:

> The "production workers" number includes workers (up through the line-supervisor [what used to be called "working-foreman"] level) engaged in fabricating, processing, assembling, inspecting, receiving, storing, handling, packing, warehousing, shipping (but not delivering), maintenance, repair, janitorial and guard services, product development, auxiliary production for plant's own use (e.g., power plant), recordkeeping, and other services closely associated with these production operations at the establishment covered by the report. Employees above the working-supervisor level are excluded from this item. (U.S. Census Bureau, 2006, pp. A-1 to A-2)
>
> *Production and related workers.* This category includes working supervisors and all nonsupervisory workers (including group leaders and trainees) engaged in fabricating, processing, assembling, inspecting, receiving, storing, handling, packing, warehousing, shipping, trucking, hauling, maintenance, repair, janitorial, guard services, product

development, auxiliary production for plant's own use (for example, power plant), record keeping, and other services closely associated with the above production operations. (Jacobs, 2005, p. 157)

The official title has become "production worker." Nevertheless, "wage-earner," the earliest designation, continues to be used (both by scholars and laypersons) synonymously with "production worker," as is the term "blue-collar worker." However, "manual worker" has gone out of fashion.

Earnings versus Wage Rates

A *wage rate* is a payment to a time-worker (generally a production worker, as discussed in Definition of Production Workers), and is expressed per unit of time; for example, dollars per hour or other time frequency. In principle, a wage rate is representative of some standard, for example, under a company-established or union-contractual-arrangement work category. *Earnings* are computed as the ratio of total payments to total time of work for a given time period—and can apply to either a given worker or a group of workers (e.g., earnings per hour [for the time period] are computed as total payments to the given worker or the group of workers divided by total worker-hours). For a given worker, earnings incorporate not only the standard (wage) payment but also payments above or below the standard (e.g., due to overtime or fines). For a group of workers, average earnings embody not only non-standard payments but also the effect of the current composition of the work force receiving the compensation.

If the goal (as it is here) is a series of average compensation received by workers, then earnings are the preferred concept. However, where earnings information is lacking, then one must have recourse to wage data.

Strictly speaking, earnings and wages (or wage rates) are distinguished as above; but, as Margo (2006l, p. 2.41, note 3) observes, even scholars are known to treat the terms as equivalent: "When series are computed by dividing total payments to labor by total time worked (for example, total hours) the convention (not always followed) is to refer to them as earnings…rather than wages." In this study, the terms "earnings" and "wages" are used in the strict sense when the distinction is relevant; otherwise, loose vocabulary may be followed (either "wage" or "earnings" denoting wages and/or earnings), for ease of exposition.

Fixed versus Current Weights

For a group of workers, earnings or wages must be weighted, in order to compute an average for the group. An unweighted average is clearly less representative than a weighted average, and a logical weighting pattern would be proportional to employment or to the labor force, that is, a given earnings or wage would be weighted by the number of workers employed, or the number engaged in the occupation, associated with that earnings or wage. Such weights may be either fixed or current. Fixed weights apply to the entire time period of the series. (In principle, a fixed weighting pattern can apply to a given segment of the series, and the fixed pattern can change with the segment.) Current weights are specific to each observation (in present context, year) of the series. Current weights incorporate a changing industry and/or occupation distribution of the work force over time, whereas fixed weights impose an unchanged distribution throughout the series (or throughout a given segment of the series). The difference between current and fixed weights is stated starkly in Rees (1960, p. 5; 1961, p. 18):

> It should be kept in mind that changes in average hourly earnings for all manufacturing reflect both changes in wage rates for particular jobs and changes in the industrial and occupational composition of manufacturing wage earners. Since the shifts in composition have on the whole been toward high-wage occupations and industries, average hourly earnings rise more than would a fixed weight index of wage rates.
>
> Our measures of average hourly earnings are affected throughout by shifts in the occupational and industrial composition of the work force, as well as by changes in wage rates for particular occupations.

The choice between current and fixed weights involves the same considerations as the choice between the earnings and wage concepts. Indeed, for a group of workers, a current-weight pattern is an inherent property of the earnings measure. (By construction, the ratio of total worker compensation to total worker-hours is current-weighted.) Where earnings data are absent or insufficient, then it is only logical to apply current weights to wage rates. As mentioned above, the objective in this study is to develop a series of compensation received by workers. Therefore shifts of workers between occupations and industries must be recognized, as workers move to enhance their earnings. Hence, ideally, the series is founded on earnings rather than wages and constructed with current rather than fixed weights. Again, just as wages can replace earnings for lack of earnings data, so fixed weights can be applied in lieu of current weights, where current-weight information is unavailable.

For completeness, argument should be provided for the opposite choice: a fixed-weight index of wage rates. Hanes (1992, p. 270; 1993, p. 733; 1996, p. 839) makes this case, where the goal is comparison of the behavior of wages in different business cycles:

> To compare cyclical patterns across time, one must have series describing the same (or arguably similar) sectors in the same way. Otherwise, one cannot be sure that differences between periods in series behavior reflect changes over time rather than differences between sectors at a point in time.... The proper measure of wages is therefore an index of straight-time wage rates paid by given firms for a set of jobs defined as specifically as possible.... The wage series introduced here indicate changes from year to year in average hourly wage rates paid for given occupations by given firms... The series were deliberately constructed as *fixed-weight* averages of wages paid for particular jobs.

However, the purpose here is different from that articulated by Hanes. The objective of the present study is to generate a series of the average hourly compensation of production workers in manufacturing. The series is worker-oriented rather than cycle-oriented. Therefore the selection of a current-weight series of earnings is warranted. To his credit, Hanes (1992, p. 270) acknowledges that, where the objective is different from his, the choice of the present study may be justified: "For some purposes one might prefer indices constructed with weights that vary to reflect changes in the distribution of the labor force or value of output across sectors; but it would be quite difficult to construct such series for periods before World War I." Notwithstanding such difficulty, and allowing for compromise and resort to proxy data, this study does develop such a series that goes back far beyond World War I.

Denomination of Series

There are two dimensions of the denomination of the desired series. First, the *time dimension* is compensation per hour. Thus the time dimension is expressed in as short an interval as empirically logical. Where component series are originally expressed per year, month, or day, they must be re-expressed per hour, making use of information on days worked per year or month and hours worked per day.

Second, the *indicator dimension* is dollars, so the full denomination is "dollars per hour." In particular, an index-number series is not acceptable. A current-weight versus fixed-weight series usually corresponds to a dollar-denominated versus index-number series; but this correspondence does not always hold. A current-weight series can readily be expressed as an index number, even though this involves a loss of information; and a fixed-weight series can be expressed in dollars per hour, even though the resulting figures are susceptible to misleading interpretation as a current-weight series.

A current-weight series denominated as dollars per hour conveys meaningful information on the level as well as movement of compensation.

Earnings versus Benefits

Compensation consists of not only earnings or wages but also benefits (as well called supplementary remuneration, supplements to compensation, or wage supplements). Earnings or wages are the monetary (cash) component of compensation, benefits are the nonmonetary component. Historically, the nonmonetary component consisted solely of in-kind benefits; these were important not in manufacturing but rather in agriculture, in the form of room and board (and statistically not readily distinguished from "pure" wages, as noted in Margo, 2006l, p. 2.41). Later, for some firms in the manufacturing sector, discounts on produced goods, subsidized meals in the company cafeteria, and in-house medical care became benefits; but these were generally not measured in a form enabling addition of benefits to wages or earnings and in any event were quantitatively unimportant relative to monetary compensation. Only in the 1930s did benefits, whether legally required or not, become consequential; and still later did benefits become quantitatively important, and increasingly so over time. Today, legally mandated benefits include social security (employer's share), workers' compensation, unemployment insurance, paid holidays and other time off; while benefits provided voluntarily by firms include life and/or medical group insurance, pension plans, company stock or stock options, supplemental unemployment benefits, severance pay, and non-mandated paid time off (but see GROSS EARNINGS VERSUS REGULAR EARNINGS).

This study uses the nomenclature average hourly earnings (AHE) to represent earnings exclusive of benefits, average hourly benefits (AHB) to signify the benefits component of compensation, and average hourly compensation (AHC) to denote total compensation. Thus:

$$AHC = AHE + AHB \tag{1}$$

Hours Concepts

Typically, hours (or average hours) are expressed per day or per week. For most of U.S. manufacturing-wage history—and certainly well into the twentieth century—the one frequency may be converted into the other by applying the, customary, six-day workweek (see, e.g., Rees, 1961, p. 36 [converts weekly to daily hours via division by six]; Adams, 1973, p. 92 [specifies 26-day month—see other references in chapter 5, *Interpolator and Extrapolator Series*]; Sokoloff and Villaflor, 1992, pp. 35, 36 [specify 310 days of annual employment, divided by 52]). Several distinctions between hours concepts are relevant.

First, there is the dichotomy between "full-time hours" and "actual-work hours." Both concepts pertain to hours at work; so full-time hours might better be termed "full-time work hours." Regarding full-time hours, Rees (1961, p. 27) writes: "Full-time hours, also called standard or prevailing hours, refer to the normal workweek of the establishment or occupation." The description of Wolman (1938, p. 5) presents more synonyms: "The full[-]time week (variously designated as normal, standard, nominal, scheduled, maximum) may be defined as the number of hours per week beyond which a shop is normally not expected to work. As such[,] the full[-]time week may be regarded as the maximum week." One may add "expected," "established," "regular" and "straight-time" hours to the list of synonyms of "full-time" hours that Rees and Wolman provide.

Actual-work hours (with synonyms work hours, plant hours, at-work hours, working hours, or workplace hours) denote the hours spent at the workplace: "Working hours are those hours for which an employee receives pay and which he spends at the employer's place of business (or elsewhere on behalf of the employer)" (Bureau of Labor Statistics [BLS], 1971, p. 11). By definition, actual-work hours are full-time hours plus net overtime hours, where net overtime hours are overtime hours minus short-time hours.

As full-time hours are a "standard" concept, they are independent of the number of workers employed. In contrast, actual-work hours for a given industry or establishment are computed as an average. The average actual-work hours per day or week is the total number of hours worked by all workers divided by the total number of workers. That is analogous to how average earnings are constructed.

Second, actual-work hours may be construed as the sum of worktime hours and non-worktime hours. Worktime hours denote the time when the workers are actually involved in the production process (i.e., "actually working"); non-worktime hours connote workplace time devoted to non-production activities: "short rest periods, coffee breaks, standby or ready time, downtime, portal-to-portal time (if paid), washup time (if paid), travel time from job site to job site within the working day, travel time away from home if it cuts across the working day, and paid training periods" (Jablonski, Kunze, and Otto, 1990, p. 18). Statistically, worktime hours typically are not distinguished from non-worktime hours. So the distinction is one of principle only.

In contrast, the third distinction is important both in principle and in statistical practice: actual-work hours versus paid hours (also called payroll hours). Paid hours are hours for which compensation (pay) is received by the workers. The difference between the two measures is paid-leave hours, time for which the worker is paid though not at the workplace. "Paid[-]leave hours include vacations, sick leave, time for union business, personal leave, jury duty, bereavement leave, and many other categories" (Northrup and Greis, 1983, p. 103).

Thus paid hours are actual-work hours plus paid-leave hours. As Rees (1960, p. 5) observes, "before 1940...there was little paid leave." Therefore, until that date, paid hours and actual-work hours are statistically indistinguishable on an average basis. From 1940, however, the difference between paid and actual-work hours can be ignored only at one's peril. Which measure is relevant depends on the purpose of the investigation. For the present study, the issue is whether AHC should be defined (and constructed) on a paid-hour or actual-work-hour foundation. *The answer is clearly actual-work-hour.* The case is made eloquently in Rees (1960, pp. 2–4):

> The conceptual difference between earnings per hour at work and per hour paid for is important only after 1939. When earnings are measured per hour at work, an increase in the time represented by paid vacations, paid sick leave, or paid holidays [for a given number of paid-hours] will increase average hourly earnings. In the BLS series [earnings per hour paid], an increase in time paid for but not worked [again for a given number of paid-hours] leaves hourly earnings unchanged. The former concept seems preferable as a measure both of the hourly income of workers and of the hourly labor costs of employers. There can be little doubt that an additional paid holiday, for example, increases both the attractiveness of a job to a worker and the cost of obtaining a given amount of work.

Gross Earnings versus Regular Earnings

Earnings can be defined in either a net or gross sense. "Regular earnings," a term preferred to "net earnings" (though neither nomenclature is in the literature), are payment for work as regularly scheduled, and for nothing else. (In the existing literature, the concept is called "straight-time earnings" or simply "earnings" or "wages" or, when nonproduction workers are included, "wages and salaries.") "Gross earnings" (also termed "gross payroll") embody all forms of direct compensation. Included in gross earnings are regular earnings plus payment for paid-leave hours plus "supplemental pay" (premium pay for work beyond the regular schedule [overtime, weekends, holidays], shift differentials, and nonproduction bonuses)—this nomenclature (though under a regular-earnings format) is used in BLS Employer Costs for Employee Compensation (see chapter 2, *National Compensation Survey and Predecessor Series*).

Regular earnings, and therefore gross earnings, are before deductions (for social security, insurance, withholding taxes, union dues, and so on). These deductions are employee contributions to various funds. Included neither in regular nor gross earnings are employer contributions for benefits. Under "regular earnings," benefits (rather than earnings) are inclusive of payment for paid-leave hours and supplemental pay. Note that, while the benefits/earnings division of total compensation is dependent on the earnings concept adopted, total compensation is unaffected. Note also that the benefits mark-up over earnings (benefits/earnings ratio) is lower for the gross-earnings than the net-earnings concept, because the numerator of the ratio is lower and the denominator higher than under net earnings.

The present study adopts the gross-earnings concept, for three reasons. First, this concept allows earnings (not benefits) to increase as a result of unpaid leave and supplemental pay. There is a certain logic in confining benefits to employer contributions to government-mandated programs on behalf of employees (legally required benefits) and to employer payments to company-based worker-beneficial programs (such as pension plans, group insurance, and savings plans). Benefits in cash (paid leave, supplemental pay) are then included with regular cash compensation, thus converting net earnings to gross earnings. Admittedly, a case on logical grounds can also be made for a regular-earnings concept. After all, any compensation other than regular earnings can legitimately be construed as a "benefit." But, second, the gross-payroll concept is in effect recommended to all federal statistical agencies by the Office of Management and Budget (see U.S. Census Bureau, 2006, p. A-2). It is interesting that, although it observed gross-payroll at one time, BLS has since turned to regular earnings (see chapter 2, *National Compensation Survey and Predecessor Series*). Third, this study is based fundamentally, at least for earnings and wages, on

Census data (see chapter 4, OPERATIONAL CHARACTERISTICS), and the U.S. Census Bureau adheres to the gross-payroll concept (see chapter 2, *Annual Survey of Manufactures*).

INDUSTRY VERSUS OCCUPATION DATA

For a manufacturing wage series, data for industries within the manufacturing sector are indicated. Such industry data, because specific to manufacturing, are the first choice. In situations (time periods) in which manufacturing data are lacking, one may resort to occupation wage data. Generation and/or usage of occupation data could be useful as an intermediate step in developing a manufacturing wage.

NATIONAL VERSUS REGIONAL SERIES

A national manufacturing compensation series, for the United States, is the objective; but, for the early period, only Northeast manufacturing data exist. Other regional information, on an occupation basis, is utilized to convert the compensation figures from the Northeast to the entire country.

SKILLED VERSUS UNSKILLED WORKERS

Where occupation data are utilized—they constitute the best wage information for conversion from the Northeast to a national basis in the nineteenth century—both skilled and unskilled workers must be represented. Unskilled workers in manufacturing would principally be laborers. Therefore *an unskilled wage alone cannot be even approximately representative of wage in manufacturing.*

SHORT-RUN VERSUS LONG-RUN SERIES

A short-run series would have the twin advantages of high degree of data reliability and consistency. However, the objective in this study is construction of a long-run, in fact the longest-run feasible, series that runs to the present and begins as early in the past as the data are found acceptable. As is discussed in detail in chapter 5, manufacturing wage data reasonably representative of the entire manufacturing sector do not exist prior to 1820. Fortunately, continuous annual data for a varying set of industries within manufacturing are available from 1800 onward. Therefore 1800 is the starting date of the compensation series. To carry this series back to 1774 or 1790 (the starting dates of some other long-run U.S. macroeconomic series) would require use exclusively of occupation and nonmanufacturing-industry data—these being the only information available—and the reliability of the series would be adversely affected.

Terminal versus Ongoing Series

Even a maximum-length series is insufficient, if the series terminates. The series should be of an ongoing nature; that is, it should be readily extendible into the foreseeable future. The property of extendibility enhances the usefulness of this long-term series.

Uniform versus Composite Series

The decisions to construct a long-run rather than short-run series and to have the series ongoing are mirrored in the adoption of a composite rather than uniform series. A series uniform in data source, data characteristics, and data reliability would be eminently desirable but cannot possibly be a long-run series. Of necessity, a long-run series is composed of a set of series, each specific to a time segment. Through adroit use of techniques in construction of the long-run series, careful selection of the ingredient series, and linking to combine the component series into the long-run series, deficiencies of the long-run series as a non-uniform series are minimized; but these deficiencies cannot be eliminated.

CHAPTER 2

Data Sources

This chapter outlines and evaluates potential sources of data for use in constructing the desired average hourly compensation (AHC) series. For average hourly earnings (AHE), earnings and wages are the fundamental data required. However, because average annual earnings (AAE) will be used as benchmark figures for part of the period, data on average number of days of operation per year (ADO) are required to convert AAE to average daily earnings (ADE), and then average daily hours (ADH) are needed to convert ADE to AHE.

Where wages rather than earnings are the concept, AHE is replaced by average hourly wages (AHW). Where both earnings and wages are involved, the symbol used is AHE.

EARNINGS AND WAGES

Bureau of Labor Statistics (and Other Department of Labor)

Earnings

<u>Current Employment Statistics Survey: Average Hourly Earnings</u> The Bureau of Labor Statistics (BLS)—created in 1884 as the Bureau of Labor under the Department of Commerce and Labor, and which received its present name in 1913 with the creation of the Department of Labor—is the obvious beginning source for earnings and wage data. Table 2.1 presents all published BLS AHE series for production workers in manufacturing. Current-weight series are discussed first, then fixed-weight series (in *Wages*). BLS current-weighted series are all weighted by current employment.

The first two entries in table 2.1 pertain to perhaps the most famous and longest-running series published by BLS: AHE. The series is certainly the longest uniform series of earnings or wages of production workers in manufacturing, and—at least from 1932

Table 2.1 Average hourly earnings or hourly wage rates,[a] production workers[b] in manufacturing[c]: BLS Series[d]

Series	Source	Period	Denomination	Hours Concept	Weights[e]
AHE (NAICS basis)	BLS website[f]	1939–	dollars per hour	paid	current employment
AHE (SIC and other bases)	Bowden (1955a, p. 802) [for 1904], BLS Web site[f]	1904, 1909, 1914, 1919–2002	"	1904–1931: actual-work 1932–2002: paid[g]	"
ECEC (NAICS basis)[h]	BLS Web site[f]	2004[i]–	"	actual-work	"
ECEC (NAICS basis)	"	2006[i,j]–	"	"	"
ECEC (SIC basis)	"	1988–2003[k]	"	"	"
ECEC (SIC basis)[h]	"	1986–2003[k]	"	"	"
EEEC	BLS (1980, pp. 319–320)	1959, 1962, 1966, 1968, 1970, 1972, 1974, 1976,[l] 1977	"	"	"
ECI (NAICS basis)	BLS Web site[f]	2001[i]–	index number	"	fixed, updated periodically
ECI (SIC basis)	"	1987–2005[i,m]	"	"	"
HEI	see table 2.2	1947–1988	"	paid	"
AHWRI[n]	*Bulletin 77* (1908, p. 7)[o]	1890–1907	"	full-time work	fixed

Notes:

[a] Excludes benefits or supplements.
[b] Also termed "production occupations," "blue-collar occupations," and "wage-earning occupations." Broader coverage where noted.
[c] Broader sectoral concept where noted.
[d] Latest revised figures. Annual, except where otherwise noted.
[e] See table 2.2 for further information about series with fixed weights.
[f] www.bls.gov.
[g] Assignment of hours concept based on information in Bowden (1955a, pp. 801, 805).
[h] All (production and nonproduction) workers.
[i] Quarterly.
[j] Begins 4Q (fourth quarter) 2006.
[k] 2002–2003: quarterly.
[l] Relates to establishments employing 20 or more workers.
[m] No figure for 1Q 1987.
[n] Manufacturing, and building and other hand and neighborhood trades.
[o] Reprinted in Rubinow (1914, p. 809), Douglas (1930, p. 76) and Bureau of the Census (1949, p. 66; 1960, p. 92; 1975, p. 169).
AHWRI = average hourly wage-rate index.
ECEC = Employer Costs for Employee Compensation.
EEEC = Employer Expenditures for Employee Compensation.
ECI = employment cost index.
HEI = hourly earnings index.
NAICS = North American Industry Classification System.
SIC = Standard Industrial Classification System.

onward—the series reflects true earnings figures. The series is part of the BLS "Current Employment Statistics" (CES) survey (sometimes termed "the establishment payroll survey" or simply "the establishment survey). BLS and state agencies collect the basic data from a sample of business establishments. In principle (but not in actuality—see the eighth limitation below), the series is consistent until the switch from the Standard Industrial Classification (SIC) system to the North American Industry Classification System (NAICS) in 2003. As a boon to economic historians, BLS carried the NAICS-based series continuously back to 1939. A comprehensive discussion of the BLS AHE series is BLS (updated, Chapter 2).

The BLS AHE is essentially a gross-earnings series:

> *Aggregate payrolls* include pay before deductions for Social Security, unemployment insurance, group insurance, withholding tax, salary reduction plans, bonds, and union dues. The payroll figures also include overtime pay, shift premiums, and payments for holidays, vacations, sick leave, and other leave made directly by the employer to employees for the pay period reported. (BLS, updated, Chapter 2, p. 3)

However, the series excludes some items that, strictly speaking, are encompassed within the gross-earnings concept: nonproduction bonuses, retroactive pay, and payment in kind:

> Payrolls exclude bonuses, commissions, and other lump-sum payments (unless earned and paid regularly each pay period or month) or other pay not earned in the pay period (such as retroactive pay). Tips and the value of free rent, fuel, meals, or other payments in kind are not included. (BLS, updated, Chapter 2, p. 3)

Positive features of the BLS AHE series are its length, its fundamental consistency over time, its monthly frequency (enhancing the representativeness of the annual figures), and the depth and breadth of its coverage (large sample of establishments, all regions of the country). However, from the standpoint of the present study, this series is beset with limitations, as follows.

First, the hours concept underlying the BLS AHE series is paid rather than work hours. The series is denominated in dollars per paid-hour, not dollars per work-hour. Unless it is susceptible to correction, the hours concept by itself excludes BLS AHE for use in constructing the desired AHE series of this study (see chapter 1, HOURS CONCEPTS).

Second, the series excludes benefits. That is an acceptable limitation; there is nothing in equation (1) or the discussion in chapter 1,

EARNINGS VERSUS BENEFITS, which prevents independent construction of the AHE and average hourly benefits (AHB) components of AHC. While earnings-and-benefits consistency in terms of source data is a desirable property of AHC, it could be that a superior series in other respects is achieved absent such consistency.

Third, the series makes other exclusions, which are less defensible. Bonuses and commissions (unless earned and paid regularly, defined as at least once a month), retroactive payments, tips, and lump-sum payments in general are all omitted from the BLS AHE series. These exclusions bias the series downward on a level basis (at a point in time); and, because of changing payments practices, the bias has increased over time:

> CES payroll data collected from employers are limited to "regular" payroll. By definition, irregular payments to employees have been excluded so as to maintain continuity of the earnings series. New compensation practices—such as paying lump-sum amounts rather than increasing basic wage rates—plus growing reliance on irregular payments such as bonuses, perfect attendance awards, cash profit sharing, etc., to reward employees are, therefore, not reflected in the CES series. To the extent that such practices replace the more traditional adjustments in wage rates, the CES earnings series become increasingly inappropriate for measuring labor cost trends in the short run. (BLS, 1986, p. 25)

For further discussion of this issue, see Abraham, Spletzer, and Stewart (1998, pp. 308–309).

Fourth, certain components of earnings that are supposed to be included in CES AHE may nevertheless be omitted by some respondent employers. Abraham, Spletzer, and Stewart (1998, p. 310) note "the failure of some CES respondents to include employees' 401(k) and flexible spending account contributions in reported earnings." This has the same effect as the mandated exclusions just discussed. The authors conclude: "In sum, there are several types of payments that are at least partially excluded from CES earnings and that have become a more important part of the earnings captured in the other earnings series since the late 1970s." They suggest that such exclusions bias CES average weekly earnings (AWE) downward by about one percentage point over the 1973–1993 period. (It is reasonable to extend their conclusion to CES AHE.)

Fifth, over time, the sampling technique evolved, albeit in the direction of improvement, beginning in the late 1940s and culminating in a comprehensive sample redesign in 2003. The "quota sampling" of the original CES survey in the 1940s was changed to probability sampling. (See Moehrle, 2001, pp. 12–13; Morisi, 2003, p. 12; BLS, updated, Chapter 2, p. 2). Clearly, this improvement reduced the sampling error over time.

Until full implementation of probability sampling, new or young establishments were underrepresented, with this underrepresentation decreasing over time as the sample size increased and sampling technique improved. Abraham, Spletzer, and Stewart (1998, pp. 303–307) provide evidence that earnings in younger establishments are lower than in older establishments, which suggests an upward bias in the level of BLS AHE in years prior to 2003. If the bias (in a multiplicative sense) were constant, then changes in the series, or the trend of the series, would be unbiased. However, the decreasing underrepresentation of young firms implies a reduced upward bias in level over time and therefore a downwardly biased trend. Abraham, Spletzer, and Stewart (1998, p. 307) believe that this bias is small: "although we suspect that the CES sample expansion had some effect on the trend in CES average weekly earnings, the effect most likely was relatively modest, perhaps no more than a percentage point, though we are unable to provide a precise quantification." (While their discussion is in terms of AWE, again it can reasonably be extended to AHE.)

Until full implementation of probability sampling, the smallest firms were also underrepresented. Rees (1960, p. 5) asserts that such firms were actually excluded: "BLS, which needs prompt and frequent reporting, uses a 'cutoff' sample that excludes the smallest firms."

Sixth, omissions in worker coverage result in a downward bias of the BLS AHE series. Barkume and Lettau (2000, p. 12) argue that the earnings of "working supervisors," a subgroup of production workers (see chapter 1, DEFINITION OF PRODUCTION WORKERS) are subject to nonreporting:

[There are] differences in how employers, who report payroll data in the CES, define "production and nonsupervisory" employees from what data users assume is the case. For example, employers are instructed to include earnings and hours data for "working supervisors," defined as workers whose supervisory duties are incidental to their job. But some employers may not include hours and earnings of working supervisors in their payroll records[,] because such employers might consider these employees as supervisors in the organization. If working supervisors generally received above-average earnings, then an employer's omission of these earnings from their statistical reports would tend to reduce reported average hourly earnings.

Seventh, the BLS AHE series is not continuous for the early years. Until 1919, figures are available only for Census years (1904, 1909, 1914), and the figure for the earliest year (1904) is not part of the official series and that for 1919 is constructed separately.

Eighth, the 1904–1931 segment of the series is inferior to, and methodologically different from, the 1932-onward segment; construction of the 1904–1931 segment is also non-uniform within that period itself. The 1932-onward segment constitutes the BLS formal, though initially pre-CES, sampling and computation of AHE. The 1904–1931 segment is discussed thoroughly in Bowden (1955a), its 1909–1931 component briefly criticized in Bureau of the Census (1960, p. 81), and the 1920–1931 component severely criticized in Rees (1960, pp. 15–19). Interestingly, the earliest (1904) figure might have the second-highest (after 1919) quality of the 1904–1931 lot. The 1904 figure is a weighted average of AHE for 17 manufacturing industries, with weights proportional to worker-hours. Bowden (1955a, p. 802) is satisfied that: "The general average appears to be closely representative of all manufacturing."

The 1909 figure has less-impressive construction. The figure uses the 1904 figure as a base and applies constructed AHE indexes for 1904–1907 and 1907–1909, these apparently derived from information only for small groups of industries and occupations. Nevertheless, Bowden's tests of the validity of the figure are positive. The 1914 figure is estimated entirely from Census 1909-to-1914 data on AWE and average weekly hours (AWH). The Census hours concept is prevailing (full-time); but Bowden (1955a, p. 804) observes: "The slight change in work schedules and the general similarity of the industrial structure in 1909 and 1914 make reasonable the assumption that average hours actually worked followed closely the trend of prevailing or scheduled hours."

The 1919 figure is the most impressive of the entire pre-1932 segment. It is an employee-hour weighted average of AHE for 27 industries, the individual-industry data emanating from BLS "extensive industry wage surveys in 1919" (Bowden, 1955a, p. 804).

The figures for 1920–1922 are the least satisfactory of the period. Bowden (1955a, p. 804) admits that the BLS industry surveys for these years "were hardly adequate for use in the historical series of hours and earnings." The main source of data is King (1923), whose figures pertain to wage-earners and salaried workers (production and nonproduction workers) combined. Rees is critical of the BLS figures because of its unsatisfactory technique of reducing King's data to those for wage-earners alone and because the King data (quarterly, from 1Q 1920 to 1Q 1922) overlap BLS AHE neither for 1919 nor for 1923, necessitating arbitrary adjustment in the attempt to achieve a continuous consistent series. Rees believes that the BLS AHE is biased upward for 1920–1922, in part because (as he, but not Bowden, states) the BLS supplemented the King data with its own estimate of AWH, which was biased downward.

For 1923–1931, the BLS made use of its surveys for 12 manufacturing industries. A 12-industry employee-hour average was computed and adjustments were made using Census prevailing-hours data (for details, see *Current Employment Statistics Survey* under DAILY, WEEKLY, OR MONTHLY HOURS OF WORK). Rees is extremely critical of the BLS figures for this time period: (1) Only 12 industries form the basis of the AHE estimates. (2) Figures for the individual industries are generally biennial and for limited payroll periods in a given year, resulting in the need for considerable interpolation. (3) With the identity AHE = AWE/AWH, AWH is biased downward and AHE biased upward (see *Current Employment Statistics Survey* again).

Ninth, and perhaps most seriously, BLS plans to terminate its AHE series in the near future. Getz (2003, p. 39) states that "the production/nonsupervisory worker [hours and earnings] series will be discontinued. BLS tentatively plans to drop these series in 2009." It is little consolation to economic historians that the date has been moved forward one year: "Planned schedule: Discontinuation of the production/nonsupervisory worker hours and earnings series is scheduled for early 2010" (BLS, 2006a, p. 2). In place of AHE will be series of average hours and earnings for all employees (production and nonproduction workers combined). The earnings concept will be improved by including irregular as well as regular payments; but the merging of figures for production and nonproduction workers without separation is unfortunate. BLS justifies its decision as follows:

> The limited scope of the production and nonsupervisory worker series makes them of limited value in analyzing economic trends. Just as important to this decision, the production and nonsupervisory worker hours and payroll data have become increasingly difficult to collect, because these categorizations are not meaningful to survey respondents. Many survey respondents report that it is not possible to tabulate their payroll records based on the production/nonsupervisory definitions. (BLS, 2006a, p. 2)

Terminating the traditional BLS AHE series is extremely unfortunate for economic historians. Further, the BLS justifications are not persuasive. First, the distinction between production workers and nonproduction workers is clear-cut conceptually (see chapter 1, DEFINITION OF PRODUCTION WORKERS). Second, the U.S. Census Bureau has no plans to drop its production-worker series, and apparently has no difficulty with the underlying data collection (see *Annual Survey of Manufactures*).

National Compensation Survey and Predecessor Series The third to sixth entries in table 2.1 relate to BLS Employer Costs for Employee Compensation (ECEC) series, which are under the rubric of the National Compensation Survey (NCS). On sampling and computation procedures for ECEC, one may consult BLS (updated, Chapter 8, pp. 1–8; 2002, pp. 1–3; 2006b, pp. 25–28).

The ECEC earnings series is unusual in that it involves a regular-earnings concept. However, paid-leave payments and supplemental pay are shown separately; therefore the earnings series may readily be converted to a gross-earnings series.

The ECEC series have several desirable properties. First, they encompass both "wages [and salaries]" (that is, earnings) and "total benefits," which together constitute "total compensation." Each series is denominated alternatively as "dollars per hour worked" and "percent of total compensation." Second, the hours concept is actual hours worked. Third, all forms of wages and salaries are included: production bonuses, incentive earnings, commission payments, and cost-of-living adjustments are in wages and salaries; nonproduction bonuses, shift differentials, and premium pay for overtime or weekend/holiday work are in benefits (along with the usual benefits).

The ECEC series also have limitations. First, they begin only in 1988 (for manufacturing production workers, called "blue-collar occupations" to 2003) and in 1986 (for manufacturing blue-collar combined with white-collar occupations). Second, the manufacturing blue-collar category ends in 2003 and is restored (and now termed "production occupations") only in 2006. Third, data are collected in only four months (March, June, September, December), and until 2002 only the March figure (considered the annual figure) is published. Fourth, until 1996 the sampling method differs from that for 1996 onward, and BLS (2002, p. 2) suggests that under the new technique, which involves more-continuous changes in the sample of establishments: "Sampling changes may have an impact on cost changes estimated over a short time interval. Consequently, BLS advises caution in interpreting short-term comparisons of costs per hour worked."

Fourth, the switch from the SIC to NAICS basis occurred in 2004, and, unlike for the BLS AHE series, the NAICS-based series is not carried back in time. So there is no SIC-NAICS overlap of the series. There is no obvious adjustment for the resulting discontinuity in the series. Fifth, BLS plans to terminate separate series for production workers not only for AHE but also for all other BLS series, including ECEC, in 2010.

Between 1959 and 1977, under the rubric of what came to be known as Employer Expenditures for Employee Compensation (EEEC), the BLS conducted nine surveys of total compensation and its components for "production and related workers" in manufacturing. The studies are summarized in the seventh entry in table 2.1. Their history is presented in Douty (1984, pp. 23–24) and BLS (1980, p. 233). The sampling and estimation procedures are discussed in BLS (1976a, pp. 175–178). A comparison with the successor survey, ECEC, is made in Wiatrowski (1999, p. 33).

Pleasing features of the EEEC are as follows. First, there is separate presentation of wages and salaries, benefits (called "supplements to wages and salaries"), and total compensation (sum of the two components). The gross-earnings concept is followed; but regular earnings (called "straight-time pay") are also shown; and, if desired, benefits can be adjusted correspondingly. Second, payments are expressed as a percent of compensation, as dollars per hour of (actual) work, and as dollars per paid-hour (termed "dollars per hour: all hours"); the middle concept is the desired one. Third, the series are for years (1959–1977) for which no other actual-work AHE series is available. Fourth—and obviously, as the EEEC survey predates the NAICS—the survey is consistent in its SIC basis.

The principal deficiency is that the years of available data are few in number and scattered. The intention of BLS was to have a biennial survey (see BLS, 1976a, p. 176), but this was performed only for the 1966–1976 period. Further, figures for 1976 are inconsistent with those for the other years, because in 1976 only large establishments were surveyed. So there is, in effect, a two-year gap between the last two usable survey years (1974 and 1977).

Additional information (beyond table 2.1) on the BLS series with a fixed (as distinct from current) weighting pattern is provided in table 2.2. All these fixed-weight series are in index-number form. Therefore, from the standpoint of this study, they have the disadvantages discussed in chapter 1, FIXED VERSUS CURRENT WEIGHTS and DENOMINATION OF SERIES. These series would not be utilized in the present study, unless alternative, current-weight, series are unavailable or unsuitable.

Table 2.2 BLS hourly earnings or wage series[a] with fixed weights, production workers in manufacturing

Series[b]	Weighting Pattern	Period	Base Period Weight	Base Period Reference	Specific Source[c]
ECI	employment	1987–1989	1980	June 1981	—
		1990–1994	"	June 1989[d]	—
		1995–2005	1990	"	—
		2006–	2002	Dec. 2005[d]	—
HEI	production hours	1947–1958	1954	1967	BLS (1976b, pp. 8–9)
	employment	1959–1963	1966	"	BLS (1976b, p. 10)
	paid hours	1964–1980	1967	"	CWD (June 1981, p. 26)
	"	1972–1988[e]	1977	1977	BLS (1985, p. 199), CWD (February 1983, p. 26; January 1988, p. 56; February 1989, p. 40)
AHWRI	wages paid	1890–1907[f]	1899	1890–1899	—

Notes:
[a] Annual, except where otherwise noted.
[b] Series are shown in table 2.1.
[c] Source for ECI and AHWRI stated in table 2.1.
[d] Series on new base also recomputed by BLS for earlier years.
[e] 1988 monthly only.
[f] Rubinow (1914, p. 795) also presents series, 1890–1903, unweighted and weighted by employment in 1899.
CWD = Current Wage Developments.

General Sources: BLS (1976b, pp. 1, 39; 1986, p. 32), Caroll (2006, p. 3), Costo (2006, p. 28), Schwenk (1985, p. 22; 1990, p. 38).

Interesting features about the Employment Cost Index (ECI), the eighth and ninth entries in table 2.1 and the first entry in table 2.2, are that it is essentially a fixed-weight analogue of the ECEC series. These two groups of series share the same sample survey, the same earnings concept, the same division of total compensation into wages-and-salaries (earnings) and benefits components, and the same hours concept. Because the weighting pattern is at so fine a level—a given occupation (job) in a specific establishment—the ECI is a fixed-weight series in ultimate form. However, as a Laspeyres index, it has the disadvantage of the fixed weighting pattern losing applicability as time goes on. To correct this disadvantage, BLS changes the weighting-pattern base year from time to time (called "rebasing"). However, the restoration of current applicability destroys the fixity property of the series over time. The ECI is not unique in involving the need to address the trade-off between current applicability and historical fixity; the BLS (or any) consumer price index entails the same issue, and BLS makes the same type of compromise for that index.

The ECI in its various aspects is discussed well in Ruser (2001). Other useful presentations are Schwenk (1985), Caroll (2006), and BLS (2000b, pp. 1–2, 120–125; updated, Chapter 8, pp. 1–8).

The Hourly Earnings Index (HEI), now defunct, is shown as the next-to-last entry in table 2.1 and the second in table 2.2. The index is discussed or summarized in Samuels (1971), Sheifer (1975, p. 4), BLS (1976b, pp. 1, 39; 1986, p. 32; 1988, p. 19), and Wood (1988). The HEI emanates from the CES survey and in fact is constructed via adjustments to the CES AHE series (or, rather, to the data underlying this series). Fluctuations in overtime-premium pay are removed, and fixed weights eliminate employment shifts among SIC three-digit industries.

However, the HEI is not invariant to shifts in employment in the SIC within a given three-digit industry (that is, among four-digit industries), and not invariant to shifts in employment among occupations in a given establishment. Also, the HEI is unusual in the differing concepts underlying its weighting pattern over time (see table 2.2). In all these respects the HEI is further from a "true" fixed-weight index than is the ECI. Periodic changes in the base period (rebasing) make interpretation of the HEI further fraught with difficulty. It is not surprising that the ECI was introduced as a superior alternative to the HEI.

Wages
From 1919 to 1935, BLS, in its *Monthly Labor Review*, presented percentage changes in wage rates in manufacturing industries. The data were obtained from the Employment and Payrolls or other predecessor to the CES sample. Table 2.1 does not have a corresponding entry, because the BLS never presented a series of these wage-rate changes (although, at a later date, Creamer, 1950, did so—see chapter 3, AVERAGE HOURLY EARNINGS OR HOURLY WAGE RATE). Creamer (1950, p. 5) observes that the reporting may be incomplete, that is, "some firms that changed wage rates may simply have failed to report." Creamer states that the series was discontinued "because of inherent deficiencies and in the belief that the

average hourly earnings series instituted in 1932 covered much the same ground." The latter is a strange belief, as the AHE is an earnings rather than wage-rate concept and is presented in tabular form. Clearly, the BLS AHE series constitutes a distinct improvement to the wage-rate-change data.

The average hourly wage-rate index (AHWRI) for 1890–1907, the last entry in tables 2.1 and 2.2, is a landmark in wage development at the Bureau of Labor (later to become BLS). For a description and discussion of the series, one may consult *Bulletin 77* (1908), Douglas (1930, pp. 73–76), Bureau of the Census (1949, p. 57; 1960, pp. 80–81; 1975, p. 152), Rees (1961, p. 6), Douty (1984, pp. 18–19), and Vangiezen and Schwenk (2001, p. 18).

Bulletin 77 (the source of AHWRI) is the culmination of a group of surveys of occupational wage rates by industry that the Bureau of Labor performed between 1900 and 1907, with data obtained back to 1890. Bureau field agents collected the information directly from employer records, suggesting a higher reliability of results (in particular, the AHWRI series) than if the information were collected via interview or mail questionnaire. The relatively large numbers involved—over 40 industries, over 300 occupations, and over 340,000 employees in 1906–1907, for example—are also suggestive of a reliable AHWRI series. Further, all regions of the United States are covered (see *Bulletin 77*, 1908, p. 17), lending credibility to the national nature of the series.

The Bureau constructed the series in three steps. First, for each industry, for each occupation a wage-rate index is computed as the average wage of all workers in the occupation. Second, for each industry, the unweighted average of the occupational wage rates is computed. Third, for total manufacturing, AHWRI is the weighted average of the individual-industry average occupational wage rates, with the weighting pattern the wage payroll of the industry in the 1900 Census (actually pertaining to the year 1899—see *Regular Census* below).

Limitations of the series are as follows. First, as observed by Douglas (1930, p. 75), the payroll weighting concept implies that higher-paid industries are given a greater weight than if the concept were employment; and the latter is the usual selection. Second, the 1904–1907 survey restricts industry coverage to industries in which payroll was at least $10 million according to the 1900 Census; this means that small industries are not just underrepresented but actually excluded. Third, at least for 1890–1903, only establishments with complete annual records for the time period are included; thus both establishments with incomplete records and new establishments are excluded.

Fourth, as Douglas (1930, pp. 73-74) states, "Since the building trades and street and sewer work were included throughout the period, the series was not strictly confined to manufacturing." Douglas neglects to mention the inclusion of "blacksmithing and horseshoeing," also considered a nonmanufacturing industry in the 1904 Census. Fifth, occupational coverage is limited to "certain important and distinctive occupations which are considered representative of each industry" (*Bulletin 77*, 1908, p. 12). The obvious implication is that unskilled labor is underrepresented. Douglas (1930, p. 73) sees this: "In general, therefore, the unskilled laborers tended to be omitted from the analysis." He observes, nevertheless, that wages rates for unskilled workers are provided for 18 of 65 industries in 1890-1903.

Fifth, there is limited coverage of pay periods within each year. Data are collected only for payroll periods considered "normal" for the establishment. Sixth, the series is a peculiar combination of time-worker and piece-worker wages, with the time-worker designation dominating. As Douglas (1930, pp. 74-75) comments:

> To secure the average *wages* per hour, the total sums of money thus built up from the hourly *rates* of the time-workers and the hourly *earnings* of the piece-workers were divided by the total number of wage-earners covered in the occupation. These results are, then, a blend of hourly rates and hourly earnings. Since, however, there were few bonuses in those days, and since many of the piece-workers necessarily had to be omitted because of lack of data [actual number of hours worked], the results are on the whole more in the nature of *rates* than of earnings.

After 1907, BLS work in the nature of series (as distinct from one-time studies) in the area of worker compensation became less impressive and remained so until BLS instituted its Employment and Payrolls (to repeat, forerunner of the CES) survey in 1932. The pertinent BLS industry surveys in the 1920-1931 subperiod are mentioned in <u>Current Employment Statistics Survey: Average Hourly Earnings</u> (eighth limitation). Only limited additional discussion is warranted here; for more information, one may consult Douglas (1930, pp. 76-83), Rees (1961, pp. 6-7), Douty (1984, pp. 19-20), and Vangiezen and Schwenk (2001, p. 18).

During 1908-1931, BLS created two sets of hourly series by industry: wages rates in specified "union" industries, and earnings in specified "payroll" industries. The number of industries in each category varied over time: 4–6 "union" manufacturing industries, and 7–12 "payroll" manufacturing industries. Thus, compared to 1890–1907, "the number of industries covered decreased sharply" Rees (1961, p. 7). For union industries, data are available annually; but for payroll industries, data are generally annual only to 1914 and biennial after that year, with some industries covered in even years and others in odd years. Thus, while the studies are systematic, the resulting series, at least for the payroll industries, show recurrent gaps. Further, the union hourly wage rates are conceptually inferior to the payroll-based AHE of the payroll industries, at least from the standpoint of the present study.

Douglas (1930, p. 78) comments: "The scope of the payroll studies was, from 1912 on, much more extensive than that covered for these industries from 1890–1907." His criterion is the number of workers included; but, as mentioned, the number of industries surveyed was much less than in the earlier period. Douglas (1930, p. 80) also praises the national coverage of the payroll series: "The establishments chosen were so selected from the various centers of the industries as to give proper representation to all sections of the country. The samples may therefore be considered to be geographically adequate." It is interesting that "no official average has ever been published of the new wage series for 1907–14" (Rees, 1961, p. 7)—and even though the BLS AHE series has observations for only three scattered years until 1919 (see table 2.1).

Bulletin 18 (1898), a Bulletin of the Department of Labor, provides average daily wage rates (termed "actual average daily wages") annually for 1870–1898 for 25 occupations in 12 cities, with the unweighted average of the city figures shown for each occupation. Wages for the years of the greenback period (1870–1878) are converted from greenbacks to gold (except for San Francisco, for which wages were originally provided in gold).

It is unknown whether the work was performed under the auspices of the Bureau of Labor or of another unit of the Department of Labor; the former is likely, as the *Bulletin* was later renamed *Bulletin of the Bureau of Labor*. The *Bulletin 18* series are described and discussed in *Bulletin 18* (1898, pp. 666–667), Long (1960, pp. 9, 13, 17, 29, 86), Douty (1984, p. 18), and Rosenbloom (1990, pp. 88–89; 1998, pp. 295–297). The entirety of the known information about the survey underlying the series is provided in *Bulletin 18* (1898, p. 666) as follows:

> [In] most instances quotations for each [occupation] were secured from at least two establishments in each city. The data are from firms

that have existed and have done business continuously since 1870, and the facts in most instances, in accordance with the rule of the Department, have been taken directly from the pay rolls...the tables...are the result of a large amount of data showing for each occupation and each year the number of employees working on full time and receiving each specified rate of pay. This information in its detail is exceedingly interesting, but almost 400 pages of the Bulletin would have been required for its publication, and for this reason only the briefer summaries of this mass of data are shown in the tables...the actual average daily wages in the 25 selected occupations from 1870 to 1898, inclusive, in each of the 12 cities in the United States...as shown by the summing up of the many quotations of individual rates paid, which were secured in most instances directly from the pay rolls of the various establishments.

In assessing the *Bulletin 18* series, one is struck by the fact that the characteristics of the series are at once both limitations and virtues, perhaps depending on one's point of view or what one wants to emphasize. First, "as to sources and methods, Bulletin 18 is virtually silent" (Long, 1960, p. 9), as the above long quotation from the *Bulletin* indicates. Rosenbloom (1990, p. 88, note 10) attempted to obtain records of the study from the Department of Labor and the National Archives, but "no trace" was found. However, the above quotation from the *Bulletin* suggests a well-constructed study, with the protocol of obtaining data directly from the payrolls of firms.

Second, the representativeness of the series has been called into question in various dimensions. Long (1960, p. 86) writes: "The number of establishments was probably small." Rosenbloom (1998, p. 297) notes "the small sample sizes on which most observations were based." Yet the Bulletin states that, for publication of the underlying data, "almost 400 pages of the Bulletin would have been required," suggesting certainly a large number of workers covered and probably also a large number of establishments. Hanes (1992, p. 274) declares: "The data themselves arouse suspicion, showing no change in nominal rates for long periods over which the Aldrich, Weeks, and Bureau of Labor reports show considerable variation for the same or similar occupations." It is true that, for many occupations, some (sometimes most) of the cities exhibit rigidity in the wage; but that is not true for national averages and likely is not true for certain regional averages as well. Two indicators of careful attention to data: some occupations have no data for one or more cities—indeed, of the 25 occupations, only seven have data for *all* cities; and, if a city is included, there are no missing observations. Also, it may be that wage rigidity does not seriously affect wage ratios between regions—and that is the role of the *Bulletin 18* series in the present study (see chapter 5, **Interpolator and Extrapolator Series**).

Confining the survey to firms that were in existence throughout the 1870–1898 period omits new firms and firms that, whether temporarily or permanently, go out of business. However, what one loses in representativeness one gains in consistency. Rosenbloom (1990, p. 88) states: "This major collection of wage data provides the only annual coverage of a consistent sample of occupations and locations from 1870 until the end of the century."

Both the occupational and geographical coverage of the *Bulletin 18* series have been criticized. Of the 25 occupations, 18 can legitimately be defined as skilled, four are "helpers" to skilled workers, and only three (two laborer categories, and teamsters) can be construed as unskilled. So unskilled workers appear to be underrepresented. However, the "laborers, other [nonstreet]" category encompasses a substantial percentage of the workforce. That occupational group corresponds to "laborers, not specified" in the Censuses of the time, which Census category generally has the largest number of workers in the nonagricultural sector. Therefore, given weighting of occupations by employment, unskilled workers would not be underrepresented.

A limitation of the occupational skilled-worker coverage is mentioned by Rosenbloom (1990, p. 89, note 11), but then he acknowledges that the limitation has little impact: "While the coverage of the data is tilted toward craft workers and away from industrial employees, the occupations included nevertheless represent a considerable fraction of the urban manufacturing labor force."

In a similar vein, the fact that the series are all occupational with no attachment made to an industry is a defect for the present study, concerned, as it is, with worker compensation in manufacturing. However, with some exercise of judgment and acceptance of the nebulous character of the result, the occupations can be associated with corresponding industries. In particular, Long (1960, p. 17) identifies ten occupations with manufacturing and six with construction ("the building trades"). The remaining occupations are readily categorized as transportation (four occupations) and "helpers" in manufacturing industries (four occupations), leaving only one unassigned occupation (street laborers).

Of the dozen cities in *Bulletin 18*, five (Baltimore, Boston, New York, Philadelphia, Pittsburgh) are in the Northeast, four (Chicago, Cincinnati, St. Louis, St. Paul) in the Midwest, two (New Orleans, Richmond) in the South, and one (San Francisco) in the West. It is easy to criticize the poor representation of the South and West. However, as Long (1960, p. 29) states, these are "large cities in all sections of the nation." In a similar vein, Rosenbloom (1990, p. 89) comments that "there is at least one city from each major region." No other wage study can make that claim for the entire 1870–1898 period.

First Annual Report of Commissioner of Labor: Results of the first survey undertaken by the newly formed Bureau of Labor appear in Commissioner of Labor (1886, pp. 91, 141–226, 295–410). The data cannot be in time-series form; for they are a cross-section for the year 1885, with average daily wage rates shown for that year alone. Yet the study is impressive, in data collection (taken by Bureau agents directly from payroll records), coverage (over 500 establishments, almost 130,000 workers), and disaggregation (presentation by occupation, industry and state; also by age-gender category: adult males, adult females, children and youths). Of the 37 industries included, only two (stone, for which see chapter 3, AVERAGE DAILY WAGE RATE; and railroad construction) are clearly not manufacturing; and there is one "miscellaneous" industry category.

The study is discussed in Douty (1984, p. 17) and Long (1960, pp. 9–10, 29), the latter observing that "the survey was based on twenty-five times as many employees as the Aldrich Report." Long (1960, pp. 145–147) provides summary tables superior to that in the study itself (Commissioner of Labor, 1886, p. 226). He shows average daily wage rates, average hourly wage rates, and average hours per day by industry, with the hourly wage-rate figures presumably calculated by Long himself. Wage rates are shown for all workers and the three age-gender categories. Long also calculates all-industry weighted averages of wage-rates, using employment weights (which he also tabulates).

Census

Annual Survey of Manufactures
The "Annual Survey of Manufactures" (ASM), conjoined with the "Economic Census—Manufacturing" (formerly called the "Census of Manufactures" [COM], here both Census names are given the rubric "Manufacturing Census" or "Census"), provides annual series of manufacturing production-worker "wages" and hours. These are not average figures but rather aggregate figures (total wages and total hours, over all workers). A potentially useful series for this study is obtained by dividing total wages by total hours, yielding AHE.

Interestingly, at one time the Census Bureau made the computation and tabulated the resulting series, which it termed "average hourly earnings of production workers" (see, e.g., Bureau of the Census, 1981, p. 1.4). Although that Census practice has lapsed, it is only logical to make the division on one's own. The resulting series may be termed the ASM AHE.

The "wages" series is actually a gross-earnings (payroll) concept. Included are "all forms of compensation, such as salaries, wages, commissions, dismissal pay, bonuses, vacation and sick leave pay, and compensation in kind" (U.S. Census Bureau, 2006, p. A-2). Wages are prior to deductions, such as employee social-security contributions, withholding taxes, group insurance, and so on. Hours are an actual-work concept, consisting of "all hours worked or paid for at the manufacturing plant, including actual overtime hours (not straight-time equivalent hours). It excludes hours paid for vacations, holidays, or sick leave when the employee is not at the establishment" (U.S. Census Bureau, 2006, p. A-2). Therefore, with its gross-earnings and actual-work-hours characteristics, the ASM AHE is particularly suited for use in the present study.

A second favorable property of the ASM AHE is that the ASM has wide coverage, with inclusion of all manufacturing establishments with one or more paid employees. A third advantage is that the ASM has always used a probability sample—another reason why small firms, as well as large firms, are given suitable representation. For details on the sampling procedure, see U.S. Census Bureau (2006, pp. C-1 to C-4) and American FactFinder Help (undated-a, pp. 2–5).

There are also some disadvantages of the ASM AHE. First, the ASM is combined with the Manufacturing Census, with the ASM occurring in years between Censuses. Census years with the figures to compute AHE are 1947, 1954, 1958, 1963, and then years ending with "2" or "7." ASM years are all the intervening years between Censuses, except for the year 1948. There are two potential problems here. (1) The intervening year 1948 lacks data to compute the ASM AHE directly. Fortunately, satisfactory indirect estimation of the figure is possible (see chapter 5, 1920–2006). (2) The ASM, though integrated with the Manufacturing Census, is a sample phenomenon; whereas the Census is a population concept, in principle a complete canvass of all manufacturing establishments. Therefore the ASM years may involve an AHE figure that is different from that which a Census would yield. However, the integration process suggests that the ASM series has a high level of consistency, because the

ASM sample of establishments emanates in large part (i.e., except for new-firm components) from the past Census itself. (There is currently a one-year lag for replacement of the former-Census information with that of the most-recent Census.)

The second limitation of the ASM AHE is that there is an inconsistency in the timing of intra-annual information used for payroll and hours. For 1947, employment is stated to be an average of the 12 monthly figures. Presumably, this statement in aggregate form applies also to hours and payroll. From 1949 to 2004, the employment figure is based on the payroll periods that include (or are nearest to) the twelfth of March, May, August, and November. From 2005 onward, these payroll periods are replaced by the March 12, June 12, September 12, and December 12 pay-periods. It is unknown how the changing procedure for incorporating intra-annual information affects the ASM AHE series.

The third limitation is the inconsistencies introduced with a switch from the SIC to NAICS basis in the 1997 Census. Inconsistencies emanate from three changes. First, some industries (prominently bakeries, candy stores where candy is made on the premises, custom tailors, makers of custom draperies, and tire retreading) entered the manufacturing sector. Second, other industries (prominently logging and portions of publishing) left manufacturing. Third, central administrative offices and auxiliary establishments (such as warehouses and research laboratories) serving a manufacturing establishment within the same organization were removed from the manufacturing sector.

The SIC-to-NAICS switch, of course, introduced an inconsistency in all dependent data series, in the year in which the change occurred. The net effect of this switch on the ASM AHE series is unknown; but something can be said. Because AHE is a ratio of two ASM series, one may legitimately presume that the effect is substantially less than that on any one related ASM series. We know only that the U.S. Census Bureau reports an approximate three-percent increase in the value of manufacturing shipments due to the switch (U.S. Census Bureau, 2002, p. v).

Regular Census

Although prior to 1947 the regular COM did not provide figures to construct total-manufacturing AHE, it did present such figures to construct AAE. Ingredients are again the numerator and denominator for a quotient. The numerator is total earnings of wage-earners. Although termed "total amount paid in wages," "total wages," or simply "wages" in various Census reports, the figure actually is an aggregate of payments to (production) workers. With the figure based on payroll data, a gross-earnings concept is involved. The denominator is the average number of wage-earners employed.

The wages and employment figures are shown in tabular form, so that the computation could be made not only at the total-manufacturing level but also at the individual-industry level. It is interesting that the Census Bureau itself never has shown the results of such computations.

The AAE interpretation of the quotient is clear: "This figure of Census average earnings can then be regarded as measuring the amounts which would be earned by the average worker who was employed for the average length of time put in by those whose names appear on the payroll" (Douglas, 1930, p. 218). This Census AAE has several advantages for the present study. First, the Census AAE (or, rather, its numerator and denominator) emanates from a complete canvass of the population of establishments rather than from a sample of the population. Second, the COM extends back over a long period of time. Third, the Census AAE readily links, at least conceptually, with the ASM AHE.

Fourth, the Census provides a clear delineation of wage-earners, with other personnel (firm members and clerks) reported, separately, for the first time in the 1890 Census, and not even enumerated in earlier Censuses (except for the possibility of being included inadvertently—see Bullock, 1899, pp. 350–351, and Long, 1960, p. 40, note 5). The 1900 Census began a detailed, separate, consideration of nonproduction workers, specifically composed of "proprietors and firm members," "salaried officers of corporations," "superintendents and managers," and "clerks and other subordinate salaried employees."

There exist limitations of the Census AAE. Fortunately, they either are not serious or can be at least partially corrected.

1. AAE is an annual concept, whereas AHE (the desired series) is hourly. This difference means that, if Census AAE figures are to be used for the present purpose, they would have to be supplemented by data on ADO for manufacturing establishments (for the conversion of AAE to ADE) and on ADH in manufacturing (for the conversion of ADE to AHE). This point is made by Long (1960, p. 40).

2. The number of workers employed according to payroll, which is the denominator of the quotient that is AAE, can involve double-counting. This would be the case for workers who shift jobs from one establishment to another *within the same payroll period*. The denominator is thereby increased and AAE underestimated. This point, made by Douglas (1930, pp. 9–10), is unlikely to be important quantitatively.

3. Census reports with the requisite information are far from continuous annually. Formally, the COM was performed every ten years from 1810 to 1900, but (a) the 1810 Census makes no mention of wages or employment, (b) there was no Census in 1830, and (c) the 1840 Census reports on employment but not wages. The Census was carried out every five years from 1904 to 1919, then biennially from 1921 to 1939. There was no Census again until 1947. It follows that the Census AAE can serve only as the basis for Census-year benchmark AHE, with other data sources required to interpolate between benchmark years and (of course) to convert AAE to AHE.

4. There is confusion regarding the proper identification of the calendar year to which the Census nominal year pertains. From 1904 onward, there is no ambiguity: by law, the 1904, 1909, 1914, and so forth, Census refers to the 1904, 1909, 1914, and so forth, calendar year. Earlier Censuses were supposed to pertain to the fiscal year ending in the Census nominal year. However, referring to the Twelfth [1900] Census, the Census Bureau (United States Census Office, 1902a, p. xvii) states: "Probably nine-tenths of our great manufacturing establishments close their books at or near the close of the calendar year, and are better prepared at that time than at any other to furnish the reports required by the Census Office. Indeed, a very large proportion of the reports made at the Twelfth Census actually related to the business of the calendar year 1899." Later, Bureau of the Census (1913, p. 18) applied this correspondence all the way back to 1850; so that the 1850 Census pertains to the calendar year 1849, and so on into the future:

> In the tables giving comparative statistics for two or more censuses in the present report it has been customary to refer to the figures in each case as relating to the calendar year next preceding the year in which the census was taken—1909, 1899, 1889, etc.—and it is probable that, as a matter of fact, at each of the censuses as far back as 1850 the major part of the totals given represent the business of the year so indicated. In the case of small establishments which do not prepare formal annual summaries of their business[,] the returns for the censuses prior to 1904 doubtless very frequently related to periods other than the calendar year[,] but large business concerns which make such formal summaries use the calendar year as their business year much more often than any other twelve-month period, and the statistics for such concerns constitute a very great proportion of the totals for all concerns combined.

Scholars who place nineteenth-century Census data in the proper year include Brissenden (1929, p. 3), Douglas (1930, p. 219), and

Rees (1961, p. 29, note 11). Rees is alone in specifically declaring: "We consider the census years to be 1889, 1899, 1904, 1909, and 1914. The original census volumes refer to the first three of these as the years 1890, 1900, and 1905, though, by the Census of 1909, the census volumes followed the practice used here in referring to earlier censuses." Unfortunately, some modern authors, incorrectly, take the Census nominal fiscal year at face value. For example, Long (1960, p. 42) states that, for Census years 1860, 1870, 1880, 1890, "data are for year ending May 31."

5. While no nineteenth-century Census is immune from criticism on quality grounds, the earlier Censuses, especially the 1820 Census, are especially susceptible to such judgment. Concerning the 1820 Census, Wright (1900, p. 27) observes: "the results are not summarized for each district, nor does the report contain any aggregate statement for the entire country—an omission due, doubtless, to the incompleteness of the returns." Sokoloff (1982, p. 288) writes: "The most serious problem with the 1820 Census of Manufactures seems to be the uneven coverage of firms... nearly all areas suffered from undernumeration of firms." He notes that small firms tended to be undercounted, and that sometimes enumerators combined figures for several small firms. Long (1960, p. 40) states: "There seemed to be agreement among the late nineteenth century critics of the census that each successive enumeration had improved in accuracy and coverage" and that "the average earnings materials have been subjected to plenty of criticism as to their meaning, completeness, representativeness, and comparability from one census to the next."

However, Long (1960, p. 40) also makes an important observation (the same point argued in *Annual Survey of Manufactures*) suggestive that the criticisms can lose much force when applied to the Census AAE figure as such, because computed AAE is a ratio of two other variables:

> It is possible that the average annual earnings were more accurate than the other data gathered by the Census of Manufactures—say, value of output or horsepower. Average earnings equal the total payroll for the census year, divided by the number of wage-earners employed; omission of an establishment would not affect average earnings as much as the total wage or total employment from which they were derived, since the errors would partially cancel out.

Further, a modern scholar considers the early manufacturing censuses to have reasonably high quality. Atack (1987, p. 291) comments: "Many scholars have regarded the 1820 census data with great skepticism.... Most of the objections, however, concern the published

'Digest.' Anyone who has worked with the original documents cannot but be impressed by the quality, detail and consistency of the information reported." Atack and Bateman (1999, p. 178) observe, regarding the 1820 Census: "Coverage of the Northeast, although incomplete, is generally thought to have been quite good, but manufacturing in the South and Midwest was seriously undercounted. Nevertheless, Kenneth Sokoloff (1982) has demonstrated that this census is capable of yielding useful information on early manufacturing...." Atack and Bateman also state (paraphrasing Robert Gallman) that even data of the 1840 Census—perhaps the worst of the lot— "deserve serious consideration once their limitations are recognized." Atack (1987, p. 291) has a sanguine view of the 1850, 1860, and 1870 Censuses: "These censuses, particularly those for 1850 and 1860, are thought to be reliable, with consistent data from state to state in a given census year." In sum, Atack and Bateman (1999, p. 181) remind us that "no single early economic data source surpasses the nineteenth-century U.S. federal census manuscripts in quality, in consistency, or in comprehensiveness."

6. There are changes over time in Census computation of the average number of workers employed, the denominator of the AAE ratio. The 1820 and 1840 Censuses simply ask for the number of persons or men employed (Wright, 1900, pp. 309–310). That instruction could be interpreted as employment at a point in time, rather than an annual average. For the 1850 to 1890 (actually 1849 to 1889) Censuses, employers are required (though the precision of instruction varies— see Atack and Bateman, 1999, pp. 179–180) to report the average number of workers for the year. For 1899 and 1904, the stipulation is for the average employment for each month. For 1909 to 1939, reported employment is for the fifteenth day of each month, or for the nearest representative day. Of course, from 1899 onward, the annual figure is the average of the monthly numbers. This history suggests several problems with the employment figure used to compute AAE.

First, the reliability of the annual employment figure is not uniform over time. Clearly, with each change, accuracy of the figure increased. However, as Douglas (1930, p. 218) notes, the magnitude of the improvement associated with the (1899) change from annual to monthly reporting is unclear: "But since some of the previous individual errors canceled each other, the increase in the accuracy of the combined figure was very much less than it was for each individual factory. Nor is it certain in which direction errors, if any, occurred."

Second, as observed by Kendrick (1961, p. 439), from (presumably) 1849 to 1889 the stated annual employment figure is not on a full-year basis: "the average number employed during the time each establishment was in operation was reported, not the average for all months of the year." The same point is made by Rees (1961, p. 31): "the average employment concept was essentially average employment during the time the plant was in operation." *If a full-year rather than time-of-plant-operation basis of employment is desired,* then as Kendrick states, there is "an overstatement of employment, chiefly in seasonal industries." Obviously, then AHE is understated.

Third, from 1909 to (presumably) 1939, there is an opposite situation. As noted by Rees (1961, pp. 30–31), seasonal industries have zero figures for months in which the plant is not in operation (reported on day 15 or nearest representative day). *If a time-of-plant-operation rather than full-year basis of employment is desired,* then employment is understated and AHE overstated.

Fourth, Rees (1961, p. 31) believes that, for 1909 and 1914 (extendible to 1919–1939), "employers probably included in their count some workers who were on the payroll on the fifteenth day of the month but were not at work or receiving pay on that day. This source of error gives us too high an average employment and too low an average daily wage." Rees' statement makes sense only if "or" is interpreted as "and." It is doubtful that this problem is one of serious magnitude.

On balance, while biases and inconsistencies are clearly associated with the employment component of AAE, their magnitude is at an acceptable level. In fact, Rees (1961, p. 31) develops evidence that the "sources of error" in the third and fourth points "are, in general, roughly offsetting."

7. The size cutoff point for inclusion in the Census changed over time. For 1820 and 1840, there was no minimum size of establishment; but, of course, this did not prevent underrepresentation of small firms (see point 5). For 1849–1899, establishments (factories and hand trades) with production valued under $500 were excluded. For 1899–1919, establishments (factories only) with value of shipments below $500 were omitted. For 1921–1939, the cutoff was $5000. "These changes in the minimum size limit have not appreciably affected the historical comparability of the census figures except for data on number of establishments." (Bureau of the Census, 1975, p. 652)

8. Factory and hand-trade establishments were combined in the 1820 to 1899 Censuses. A fundamental change in coverage occurred with the 1904 Census. Hand (and custom) trades were omitted, and only factory establishments included. The 1904 Census provides data

for factory establishments by industry in 1899; so 1899 is the only year for which data for hand trades and factories are available separately. Easterlin (1957, pp. 640–641) offers an excellent discussion of the changeover and the sometimes arbitrary decisions of the Census Bureau regarding inclusion or exclusion of a "neighborhood establishment" (the criterion the Bureau used to distinguish a hand trade from a factory—see chapter 1, DEFINITION OF MANUFACTURING).

The shift in coverage from both hand trades and factories to factories alone is an important change, and cannot be ignored in any use of data spanning the 1899 Census.

9. The 1899 Census stipulated that hand trades not carried on in a shop were to be excluded. Home production is that type of hand trade. Easterlin (1957, p. 640) observes: "That some home production was included in the preceding censuses seems probable, though limitation of the canvass to establishments with products valued at $500 or more probably restricted the coverage." The technical inconsistency is not serious, just as it is not for the change in the size cutoff (point 7); but data on home production in the 1920s "suggest that the volume of manufacturing activity omitted, by the exclusion of homework from 1899 onward, is not negligible." (Easterlin, 1957, p. 640)

10. An inconsistency over time that cannot be ignored is the classification of industries in manufacturing rather than in other economic sectors. Any work on the manufacturing sector that transcends more than one Census year must make adjustments for changes in industry classifications. The issue is discussed thoroughly in Easterlin (1957, pp. 638–639, 646), and is treated also in Long (1960, pp. 40–41), Fabricant (1942, pp. 206–214), and Rees (1961, p. 29). The relationship to each alternative sector is treated briefly here.

Agriculture: The 1849 and 1859 Censuses include agricultural-processing industries that are not included in other years.

Forestry: Logging operations, previously excluded from the lumber industry (an industry within manufacturing), are included in that industry from 1904 onward.

Mining: Certain processing operations associated with mining are included in the 1849 and 1859 Censuses but not in others.

Construction: Bridge-building is an industry in the 1849 and 1859 Censuses but not thereafter. Many construction trades are included in the 1849–1899 Censuses, but not afterward.

Transportation and Public Utilities: The manufactured-gas industry and railroad repair shops are included in the 1849–1935 Censuses (except for 1879).

Trade: Easterlin (1957, p. 639) observes that wholesaling or retailing functions carried on by manufactures are supposed to be excluded from the Censuses; "in the earlier years, however, they may have been included to some extent."

Services: Dentistry is an industry in the 1859–1889 Censuses, and laundry-work in 1859.

Special Reports

Man-Hour Statistics for Selected Industries: In the 1930s, the Census Bureau, in conjunction with BLS, provided reports on "Man-Hour Statistics" for selected industries for the years 1933, 1935, 1937, and 1939. AHE and related information for wage-earners are shown for a large, and increasing, number of industries: 32 in 1933, 59 in 1935, 105 in 1937, and 171 in 1939. The studies are Bureau of the Census (1935, 1938, 1939, 1942), and they are discussed in Rees (1960, pp. 7–9) and Jones (1963, p. 377). The industries are classified more finely than in the regular Census.

Advantages of these special reports are the gross-earnings basis, the actual-work hours concept, inclusion of both full-time and part-time workers, the generally high coverage of establishments and employment within the industries, inclusion of firms of all sizes, and computation of AHE as the average of 12 monthly figures. Limitations are the sample rather than population coverage (a defect only compared to the full Census) and the fact that the figure for AHE over all industries is not consistent over time (in view of the changing, and systematically increasing, sample). Also, there is no randomness to the sample. Industries are "selected"—"the industry-sample coverage of the Census did not provide data from a spectrum of all categories of manufacturing industries" (Jones, 1963, p. 377)—and establishments with outlying observations are excluded.

Bulletin 93: A special investigation of the 1904 Census, *Bulletin 93* (1908), collected and processed weekly earnings for wage-earners in manufacturing. Brissenden (1929, pp. 287–291) discusses the investigation, revealing that it was supervised by W. M. Steuart, Chief Statistician for Manufacturing in the Census Bureau.

Bulletin 93 has several praiseworthy features. First, the data pertain to earnings rather than wage rates; and the inquiry is emphatic on this point: "The distribution of the employees must be made according to actual earnings, not rates of pay" (*Bulletin 93*, 1908, p. 9). Second, although the goal of incorporation of the full census of establishments was not attained (mainly for reason of unsatisfactory returns), coverage is high: about 63 percent of establishments and half of wage-earners in manufacturing. Third, there is an excellent summary table of results, showing average weekly earnings for all wage-earners and for adult men, adult women, and children separately. The average for each of the four groupings emanates from a tabulated frequency distribution of the number of wage-earners according to 13 weekly earnings categories (less than $3, $3 to $4, etc). Fourth, the reliability of the data appears to be high, with payroll records the information source of choice. "Verbal statements were accepted only when there were less than 10 wage-earners reported by the establishment" (*Bulletin 93*, 1908, p. 10).

One problem with *Bulletin 93* is its restriction to a single year, reducing its usefulness for understanding the movement of earnings over time. A second limitation is the nature of the information sought and recorded. For each establishment, the data pertain to "the week in which the largest number was employed during the year... For some establishments it was difficult to obtain a report for the week during which the largest number of wage-earners was employed, and the report was prepared for a representative week" (*Bulletin 93*, 1908, pp. 9–10). Brissenden (1929, p. 288) interprets the two instructions in an integrated way: "what was done in this special investigation was to secure, for a representative week of full operation in the industry concerned, a record of actual earnings received by the wage earners of each establishment."

One consequence is that the data for the reporting establishments do not pertain to the same week. A second consequence is that the data are an approximation of full-time earnings in some sense, although Brissenden (1929, p. 291) argues against that conclusion, stating "definitely that the money sums reported as average weekly earnings in 1904 are generally somewhat less than full-time earnings." In any event, the results of the report are difficult to interpret and to incorporate in a study going beyond that one year.

Dewey Report: The special Census report by Davis R. Dewey (1903), a prominent academic economist and statistician, provides wage data for 34 selected industries in manufacturing for 1890 and 1900. Both hourly and weekly wage rates were collected. The study was deliberately restricted "to a few stable and normal industries" (Dewey, 1903, p. xiv). Traditionally, the Dewey Report has been both praised and criticized—by the same authors, namely, Abbott (1905, pp. 333–336), Coombs (1926, p. 25), and Brissenden (1929, pp. 260–262). On balance, their judgment is uniformly favorable:

"While the Dewey Report may fall short of the ideal that should be set in the matter of wage statistics, it is, nevertheless, the best report of the kind that has ever been published."—Abbott (1905, pp. 335–336)

"It was carried on under the best principles of any wage investigation up to that point."—Coombs (1926, p. 25)

"On the whole, the Dewey report was probably the most important and most reliable report on wages which had, up to that time, been published in the United States."—Brissenden (1929, p. 261), quoted approvingly by Long (1960, p. 10)

Positive features of the Dewey report include the careful selection of industries to incorporate all important branches of manufacturing, the large number of establishments (over 700) and workers (over 100,000), and the incorporation of "all occupations in the establishments sampled," the last quoting Rees (1961, p. 40). Also, the report is amazingly detailed in every way conceivable. Further, the summary, such as it is, compares results for 1890 and 1900 via the median and quartiles. Certainly, presenting a distribution of wages for each category provides more information than if simply the, customary, mean was shown. As Coombs (1926, p. 25) comments: "there is not the excessive—and sometimes false—appearance of accuracy that a mean would give."

Some criticisms of the report are obvious and perhaps unfair. Data were collected for only two years (the stipulated scope of the inquiry); the compensation concept is wage rates rather than earnings (an explicit choice); within each industry, only a small proportion of the establishments were sampled (nevertheless more establishments than in the typical contemporary non-Census wage investigation). Also, Rees (1961, p. 40) observes that "it is almost uniformly true of such nonrandom samples of wage data that they overweight large or high-wage firms and are, therefore, somewhat biased upward" (random samples are not to be found in nineteenth-century investigations).

Other criticisms are more to the mark. There is no summary as such. While there are many tables, "the tables are too detailed to serve the purpose of a summary" (Abbott, 1905, p. 334). Brissenden (1929, p. 261) makes the point with vigor:

> It is significant that Mr. Dewey did not see fit to make any consolidation of his results for the different industries. Not only did he feel, apparently, that he was not warranted in consolidating the 33 industries to show a final frequency distribution for all industries combined and for the United States as a whole, but he did not even consolidate all occupations, sex and age groups, within the different industries.

In this light Long (1960, pp. 148–149) provides a great service in summarizing the 1890 results of the Dewey Report by industry and for the all-industry total, for various age-sex categories as well as for all males, all females, and all workers. Unfortunately, though understandable in terms of the period (1860–1890) of Long's study, Long did not make these computations also for the year 1900—nor, to my knowledge, has any other scholar done so.

Rees (1960, pp. 39–40) sees two defects of Long's summaries. First, because the underlying wage distributions tend to be skewed to the right, their medians are below their means; therefore Long's summary estimates are biased downward. Of course, the use of the median instead of the mean to measure central tendency emanates from the original, Dewey, study—Long had no choice. Second, in weighting industry medians by the Dewey-Report employment, high-wage industries are overweighted. Rees comments that a superior source of employment weights would be the COM itself. One may comment that the two biases are in opposite directions.

Weeks Report: One cannot dispute the judgment of Coehlo and Shepherd (1976, p. 206): "The Weeks Report is one of the most important sources and more complete collections of price and wage data for the United States in the 19th century." The study, Weeks (1886), was conducted by Joseph D. Weeks, "expert and special agent" (Weeks, 1886, p. iii) of the Census Office, in connection with the tenth (1880) Census. The Report is described and critiqued in Abbott (1905, pp. 332–333), Brissenden (1929, p. 257), Long (1960, pp. 7–9, 12–13), Coelho and Shepherd (1976), and Margo (1992, p. 177; 2000b, pp. 7–8; 2006l, p. 2.43).

The data in the Weeks Report emanate from payroll records. Almost all the wage-rate observations are per day; and the hours concept is straight-time. Long (1960, p. 11) acknowledges the objective but questions the implementation. "while the Weeks Report attempted to eliminate overtime and other premium payments so as to express wages in the price of a regular workday, and to take account of allowances and deductions so as to make the daily wage reflect the actual rather than the nominal wage, it could scarcely do so adequately." It is not clear that Long's skepticism is warranted.

The Report includes 53 industries; 46 can arguably be classified as manufacturing, 2 as construction, and 5 as mining. A total of 627 establishments are included. Coverage is broad, but by no means complete. Weeks (1886, p. ix) states:

> It will be noticed that a few quite important industries are not represented in this report, while the rates of wages in others are not as full as could be desired. For this the special agent is not accountable. No

important branch of manufacture was overlooked. Schedules were sent to the most prominent establishments in all of the chief industries, and if some are not represented it is because returns of sufficient detail or exactness could not be obtained after repeated solicitation.

Within each industry, for each establishment, for each occupation within the establishment, the average wage rate is shown annually. In principle, the time period is 1830–1880. However, most series begin after 1850, and few before 1840. It is fair to say that, from a national standpoint, very little of the antebellum period is covered adequately. Also, to use Robert A. Margo's terminology, the sample is "retrospective." The Report consists of data only for firms that were in existence at the time of the survey. Firms that existed at one time but later went out of business are not considered.

The retrospective property has two negative consequences. First, the sparseness of antebellum data is a consequence of omitting firms that existed in the antebellum period but then failed. Second, the sample is clearly not random. A positive consequence is the continuity of the wage series—same firms, same occupations, over a long period of time.

An unambiguous negative characteristic of the Weeks Report is that no information is given about the (presumably employment) weights by means of which the average wages are computed. In fact, neither total-establishment (total-firm) employment nor intrafirm individual occupational employment is provided. Therefore researchers, if they wish to use the Weeks series above the individual establishment-occupation cells (e.g., to compute series for total manufacturing), must either weight establishments equally or use externally obtained weights (such as regular-Census data on employment by occupation or industry).

Related to this problem is the fact that Weeks does not summarize the results for wages. An early critic stated the situation well: "No summary of any kind is attempted, and the data remain a vast array of facts which have never been averaged or classified to make them throw any light whatever on the question of wages" (Abbott, 1905, p. 333). Later researchers (e.g., Lebergott, 1964; Long, 1960; and Coelho and Shepherd, 1976) did make efforts to correct the situation.

The geographic coverage of the Weeks Report is good in the 1860–1880 period, with all regions represented with reasonable, though varying, adequacy. This is certainly the judgment of Lebergott (1964, pp. 297, 303), who writes: "the Weeks reports...[are] both numerous and widely spread geographically" and "the Weeks data...

report [wage] trends in many states in addition to the Northeastern bloc." However, there are geographical limitations even for that period. Two are noted by Coelho and Shepherd (1976, p. 204, note 4). First, all establishments in the Report are from urban areas. Second, during the Civil War, for the East South Central region, only establishments in cities and towns under Union control are included.

The situation is different in the antebellum period, for which there is little coverage of the South and none of the Mountain and Pacific regions (see Coelho and Shepherd, 1976, p. 227). As Margo (2000b, p. 8) observes, "the antebellum data in either [Weeks or Aldrich] report pertain almost solely to the Northeast before 1850."

Social Statistics: Daily wage data were collected by the "Census of Social Statistics" for the 1850, 1860, and 1870 Censuses. The only worker categories of conceivable interest, for the present study, are nonfarm day laborers and carpenters. Data were published for 1850 and 1860, but not for 1870, in the form of individual-state averages for each occupational category (DeBow, 1854, p. 164; Secretary of the Interior, 1866, p. 512), but with no national averages presented. The published wage data were used by Lebergott (1964, p. 271). The intrastate average daily wage (ADW) data in the manuscripts of the Census of Social Statistics were rediscovered by Margo (1998, p. 52; 2000b, pp. 30–31).

The Census states that "The [wage and other] information required in this [Social Statistics] schedule is not to be ascertained entirely by personal inquiry of individuals, but in part from the public records and reports, and public offices of towns, counties, States, or other sources of information" (DeBow, 1853, p. xxiv; Census Office, 1860, p. 28). Margo (2000b, p. 30) comments, wisely, that "it is highly doubtful that such records would provide the necessary wage evidence, and it is reasonable that marshals obtained the great bulk of quotations from 'personal inquiry of individuals.'"

Oral communication rather than written record casts doubt on the reliability of these, and any, data. As Long (1960, p. 12) comments, in another context: "data...based...on actual business records,...while subject to clerical or other technical error, do not depend for their accuracy on the memory, truthfulness, or knowledge of an employee or some member of his family."

Congress and Treasury

Aldrich Report: The Aldrich Report (1893) is named for Nelson W. Aldrich, chairman of the Senate Committee on Finance. The data in the Report were collected by the Department of Labor, under the direction of Commissioner Caroll D. Wright. Impressive for the time, the data are not only presented essentially in raw form, but also artfully summarized in tables and analyzed by Roland P. Falkner, an academic statistician hired by the Committee (see chapter 3, AVERAGE DAILY WAGE RATE). The Report is discussed and assessed in Abbott (1905, pp. 339–340), Coombs (1926, pp. 26–27), Long (1960, pp. 6–7), Lebergott (1964, pp. 289–295, 303–304), Douty (1984, pp. 17–18), Hanes (1992, p. 274), and Margo (1992, pp. 176–177; 2000b, pp. 7–8).

Wages per day are presented by firm in the form of averages for each occupation covered. Firms, nearly 100 in total, are arranged by industry and by state; data are taken from payrolls of the firms, and two payroll periods (January and July) are presented for each year. Impressively, and in sharp contrast to the Weeks Report, employment by occupation is shown for each payroll period for each firm. The time period is supposed to be 1840–1891; but only 61 wage series begin in 1840. For 1860–1891, data are temporally complete for a total of 543 wage series in 21 industries. However, of the 21 industries, only 12 are truly manufacturing and not double-counted (see table 3.2, note e).

Some observers praise the Report. Abbott (1905, p. 339) considers it "our one great collection of wage statistics" and "an invaluable source of information to students of economics and statistics." Coombs (1926, p. 26) calls it "the most useful collection of wage statistics of the last century." Douty (1984, p. 18) states: "Despite its many limitations, the Bureau's work for the Aldrich Committee is the major source of information on the structure and course of wages in this country from 1860 to 1890, and yields some insights for the years back to 1840." Lebergott (1964, p. 289) has the opposite view, describing both the Aldrich and Weeks reports as providing "a haphazard collection of wage data that need analysis, allowance for bias, and careful weighting before they can be utilized." Also, Lebergott is extremely critical of the Aldrich data as compiled by Long (see chapter 3, AVERAGE DAILY WAGE RATE).

Certainly, the problems of the Aldrich data are clear. First, "the industries varied widely in their importance in the economy" (Long, 1960, p. 7). Second, for each industry, the sample of firms is small. Third, the occupations covered within a firm can be far from comprehensive; this is particularly true for textile industries (Lebergott, 1964, p. 294). It is also apparent that, for most if not all industries, unskilled workers receive less coverage than skilled workers. Fourth, just as for the Weeks report, the sample of firms covered is retrospective (see Weeks Report). Fifth, the antebellum period receives scant coverage. Sixth, the Report is largely confined to the Northeast (New England and Middle Atlantic regions).

Young Report: A collection of wage rates for the year 1869 was assembled and authored by Edward Young (1871, pp. 202–215), Chief of the Bureau of Statistics of the Treasury Department. The data are presented as average weekly wages (sometimes "wages or earnings") by occupation and state for a small number of manufacturing industries; for a large number of "miscellaneous occupations" within manufacturing, nationally only; and for "mechanical labor" occupations, some nonmanufacturing, by state with regional groupings. Lebergott (1964, pp. 263, 272) praises the author, though he corrects some of the figures. The single year of information and the haphazard collection of industries detract from using the Young data in the present study.

McLane Report: Unusual for its time, an "enormous study" of manufacturing wage rates was performed by the Treasury Department ostensibly for 1832 (though the data pertain principally to the year 1831—see chapter 5, *Average Annual Earnings*). The study is called the McLane Report (1833), after Treasury Secretary Louis McLane. The above quotation is from Lebergott (1964, p. 285), who goes on to write: "It [the McLane Report] provides estimates of earnings in thousands of firms and hundreds of individual towns by such hand tradesmen as house carpenters, saddlers, blacksmiths, etc. In a real sense such comprehensive materials have not been published since." The Report consists of over 1900 pages in two volumes. Information was obtained by Treasury enumerators who visited the establishments. The Report shows average daily, weekly, or monthly wages, by occupation for firms or towns. The Report is assessed in Lebergott (1964, p. 285), Chandler (1977, pp. 60–62), Sokoloff (1982, pp. 287, 292–293), and Goldin and Sokoloff (1982, p. 745, note 6).

One limitation of the Report reflects the fact that, even though it may have been a substitute for the absence of manufacturing data from the 1830 Census, the Report is not a census. This is seen in two respects. Geographically, the Report is restricted to the Northeast. The only exception is "a short and very incomplete statement on Ohio" (Chandler, 1977, p. 61). Within the Northeast, "large establishments and firms located in New England (particularly Massachusetts) are overrepresented in the McLane Report" (Sokoloff, 1982, p. 292). In particular, as noted by Sokoloff, coverage is especially deficient for New York and absent for western Pennsylvania.

Also, small manufacturers in general are underrepresented, except for Massachusetts.

Another deficiency is the overrepresentation of the shoe industry in Massachusetts, with Lebergott (1964, p. 285) commenting on "the enormous number of firms reported."

A third problem with the Report is the nature of its presentation of data. There are no summary averages of any kind—not by occupation, not by industry, not by state, and not for total manufacturing by state or for all covered states together.

State Labor Bureaus

<u>Massachusetts:</u> Thanks to the leadership of Carroll D. Wright, then chief of the Massachusetts Bureau of Statistics of Labor, a massive amount of data on wages in that state is assembled in Wright (1885). The information was obtained from "books of account" (Wright, 1885, p. 41), that is, payroll records, of firms. Average wage rates are almost all daily and for males; they are arranged by occupation annually for 1752–1860. Even though confined to one state, the study is distinguished by the mid-eighteenth-century beginning, the large body of original data (over 4,600 statements containing over 9,000 wage quotations), and by the inclusion of both skilled and unskilled occupations. David and Solar (1977, pp. 61–62) and Lindert and Williamson (1982, pp. 420–421) find the study praiseworthy. Margo (2000b, p. 8) is reserved, observing that the survey is retrospective in nature, that wage quotations are not continuous annually for any occupational category, and that (like the Weeks Report) there is no information on the number of workers underlying each wage observation. Also, David and Solar (1977, p. 61) note that "considerable obscurity surrounds the number and distribution of establishments contributing the wage quotations."

<u>Various States:</u> Beginning in the 1880s, some state labor bureaus generated contemporary series of ADW or AWE. After World War I, over time, as BLS intensified its data collection, the states reduced or eliminated their own activities. Most interesting for this study, and analogous to the computation of AAE at the national level using Census data (see *Regular Census*), bureaus collected state data on total earnings and average number of workers, the ratio of which yields individual-state AAE. Unlike the Census, the state bureaus made heavy use of questionnaires rather than field agents and conducted only a sample of establishments; both features suggest a reduced reliability of the state statistics. The data of some states have

specific deficiencies (inclusion of salaried employees, ambiguities of statistical technique). However, beginning in 1914 (New York) and the 1920s (certain other states), the components of AAE were collected continuously monthly rather than discontinuously annually—in this respect, certainly an improvement over the Census methodology. The experience of the state bureaus is discussed in Brissenden (1929, p. 267), Douglas (1930, pp. 220–222, 233–234), Rees (1961, pp. 131–136), Douty (1984, p. 27, note 8), and Hanes (1992, p. 275)

Individual-state series have obvious geographical restriction, but they can be combined to approximate, or at least interpolate, a national average-earnings series. One author has done just that (see chapter 3, COMPOSITE SERIES).

Antebellum Records of Firms

For the antebellum period, wage information is scarce. Sources other than government can help to fill gaps in data requirements for this study. One such source is the records of contemporary businesses. These records could be receipts, payrolls, bills, account books, and the like. It is possible that undiscovered archives of firms and even attics of houses contain a trove of antebellum wage information. Some antebellum records that have been mined pass the test of covering both skilled and unskilled workers, and they are considered here. (The study of Robert G. Layer is excluded, because it incorporates only skilled workers—see Layer, 1955, p. 12, note 8.)

Manufacturing Firms in Brandywine Region: The Brandywine Region, in Southeastern Pennsylvania and Northern Delaware, was an early location for manufacturing, partly because of the existence of waterpower, partly because of the proximity to Philadelphia. The records of three firms in the area—the gunpowder firm E. I. DuPont; the textile firm Charles I. DuPont; the textile firm Bancroft, Simpson and Eddystone—were discovered by Donald R. Adams Jr., and described in Adams (1982, pp. 905, 916–197) and Margo (2000b, p. 13). E. I. DuPont supplies wage and earnings information for the time period 1802–1860, while the other two companies provide data for subperiods. Wages of male workers are distinguished from those of female workers. On Adams' compilations of these data, see chapter 5, *Interpolator and Extrapolator Series*.

Iron Firms in Eastern Pennsylvania: Jeffrey F. Zabler (1972, p. 110) discovered "manuscript data on wages for a very wide sample of iron-producing firms in eastern Pennsylvania for the 1800–1830 period." Wages are monthly, and several occupations in both the skilled and unskilled categories are distinguished. The nature of the industry indicates that the wages are for males. Zabler's series are discussed in chapter 5, *Interpolator and Extrapolator Series*.

Maintenance of Erie Canal: The Erie Canal Papers, in particular, "check-rolls and workmen's receipts," are the source of wage rates for skilled and unskilled occupations for maintenance work on the Canal. The discovery and compilation of these data are due to Walter B. Smith, who describes the data in Smith (1963, pp. 298–299, 301). Except for cooks, all workers on the Canal were male. The Canal was completed in 1825; the data are for 1828–1881. There are a total of about 30,000 wage-rate observations, approximately 90 percent of which are for common labor. Margo (2000b, p. 9) judges that Smith is "perhaps the most famous such study [of antebellum wage movement based on archival records]." Margo's (2000b, p. 10) one criticism is that Smith's selection of the mode as the measure of central tendency "might impart a spurious stability to nominal wages"; but Smith defends the choice over the mean.

Early Philadelphia: Donald R. Adams Jr. discovered archival material (account books, ledgers, individual wage bills, and receipts) containing an impressive array of wage-rate data in Philadelphia for the time period 1785–1830. The most important source is the papers of Stephen Girard. "Girard, a Philadelphia financier and philanthropist, maintained meticulous records of business dealings, and his records have yielded an abundant collection of wage quotations" (Margo, 2000b, p. 11). For discussion of the material, one may consult Adams (1967, pp. 4–10, 196–200; 1968, pp. 404–405).

Records of Civilian Employees of U.S. Army

A unique set of antebellum wage information was discovered by Robert A. Margo and Georgia C. Villaflor. *Report of Persons and Articles Hired* is a monthly report prepared by the quartermaster at a military facility (army post or naval yard), and includes payroll records of civilians employed. For each worker (almost all male) at the specific location, the occupation, daily or monthly wage, and number of days worked during the month are shown. The great bulk of the existing *Reports* pertain to Army forts. *Reports* were found for the years 1818–1905. Interestingly, what is extant are not the original *Reports*, but rather duplicates, stored at the National Archives. Characteristics of this intriguing data source are described in Margo and Villaflor (1987, pp. 874–878) and Margo (1992, pp. 181–182; 1998, pp. 52–53; 2000b, pp. 25–34). Margo's resulting series are discussed in chapter 5, *Interpolator and Extrapolator Series*.

Private Surveys

The most important private survey of manufacturing wages or earnings is that of the National Industrial Conference Board (NICB). For the time period 1914–1948, based on a mail questionnaire, the NICB published an annual (and monthly) AHE series for production workers in manufacturing. Component industries—originally 27, reduced to 25 in 1920—have their individual AHE computed as the ratio of total payroll to total number of hours worked. A gross-earnings concept and actual-work hours are applied. The NICB annual all-manufacturing series is tabulated in full in NICB (1950, p. 336), Bureau of the Census (1960, p. 94; 1975, p. 172), and Margo (2006h). The series is described and assessed in Coombs (1926, p. 33), Beney (1936, pp. 15–23), NICB (1950, pp. 336–337), Bowden (1955b, pp. 920–921), Rees (1960, pp. 15–16), Bureau of the Census (1960, p. 82; 1975, p. 154), David and Solar (1977, p. 64), and Margo (2006h).

Because of its actual-hours basis, its construction analogous to Census-based AAE, and its monthly frequency, the NICB series is worthy of serious consideration for use in the present study. However, the limitations of the series must be recognized. Two are not serious. The number of industries is really 21 rather than 25, as "foundries and machine shops" counts as five industries ("foundries," shown separately, and four other branches). The 1914–1948 period is not continuous. The only 1914 figure is for July; and 1920, 1922, 1948 annual figures are averages of seven, six, seven months.

Important weaknesses in the NICB series emanate from the fact that it is not based on a Census. First, the survey is not a random sample with mandatory reporting; rather, the survey is dependent on voluntary reporting on the part of cooperating firms. Second, large firms are overrepresented in the sample of most industries, which suggests an upward bias in average earnings. Third, firms in Southern states are grossly underrepresented. Fourth, the individual-industry AHE series are combined into an all-manufacturing series via a fixed weighting pattern—Census employment in 1923. Thus the NICB series violates the argument that a current-weighting pattern is preferred (see chapter 1, FIXED VERSUS CURRENT WEIGHTS).

A less-extensive private survey is a one-time study of the National Bureau of Economic Research. The survey is described, and the pertinent results presented, in King (1923, pp. 9–21, 113); it is criticized in Barnett (1924, pp. 109–110) and Rees (1960, pp. 17–18). Good features are the adoption of payroll (that is, gross) earnings and actual-work hours, the application of the usual AHE computation, the separate results for (and strict definition of) manufacturing ("all factories"), and the assembling of quarterly data. Limitations are the use of questionnaires, the shortness of the series (only the eight quarters of 1920–1921 and the first quarter of 1922), and the combining of production and nonproduction workers. Rees (1960, p. 18) comments that "in a period of severe depression [characteristic of the period covered by the survey], wages undoubtedly differed from salaries in movement as well as in level."

Days of Operation

Census: Data on ADO in manufacturing are needed to transform AAE to ADE. The 1860 Census was the first with a question on months of operation (for plants "idled for long"); but, according to Atack and Bateman (1999, p. 180), "such reporting was rare." The 1870 Census collected establishment data on the number of months of operation, "reducing part time to full time" (Wright, 1900, p. 314). The 1880, 1890, and 1900 Censuses did the same with explicit instructions to report the number of months that were full-time, three-quarters time, two-thirds time (1880 only), half-time, one-quarter time (1890 and 1900), and idle (see Wright, 1900, pp. 314, 315, 363; U.S. Census Office, 1902a, pp. cxxxiii–cxxxiv). None of these Censuses tabulates the results, or provides any information whatsoever on months of operation. However, consulting manuscript schedules, Atack, Bateman, and Margo (2002), preceded by Atack and Bateman (1995), do so for the 1870 and 1880 Censuses (see chapter 5, *Days of Operation*).

Only the 1904 Census tabulates number of days of operation. This Census shows a frequency distribution of the number of establishments according to the number of days of operation, for the United States and also by individual state (Bureau of the Census, 1907, pp. 542–543). There are ten equally spaced intervals, ranging from "31 to 60" to "301 to 330" days, and also end-intervals "30 and less" and "331 to 366" days. No mean or other measure of central tendency is computed.

State Labor Bureaus: Data on number of days of operation of manufacturing within a state are assembled by individual-state labor bureaus. Rees (1961, pp. 131–136) provides an excellent summary of these data, with sources, for the time period 1890–1914, extendible to 1919. The states with usable data for the period are Massachusetts, New Jersey, and Pennsylvania. Data for other states are inconsistent over time, cover only a small number of industries, or are available only for intermittent time periods.

DAILY, WEEKLY, OR MONTHLY HOURS OF WORK

Data on ADH of work are required to convert ADE into AHE. Many work-hours series are on a weekly or (rarely) monthly rather than daily basis. The weekly frequency is acceptable for our purpose, because, for the nineteenth century and well into the twentieth century, the typical workweek in the country was six days per week (see chapter 1, HOURS CONCEPTS). Therefore AWH are readily transformed to ADH, simply by dividing by six; and sources of weekly or of daily hours are considered together in this section. Monthly hours can also be so converted, again with minimum scope for judgment on the number of working days per month. In general, the format of EARNINGS AND WAGES is followed; because most hours data are associated with earnings or wage series, in the same source.

While, as a general rule, actual-hours worked are preferred to normal hours, it happens that, given data availability, full-time hours are utilized for the construction of the pre-1920 segment of the new AHE series. Fortunately, there is a certain logic to this use (see chapter 5, *Average Daily Hours*). Thus data sources of full-time hours cannot be neglected. However, data on a paid-hours basis could be usable only with adjustment.

Bureau of Labor Statistics (and Other Department of Labor)

Current Employment Statistics Survey
Connected with the BLS AHE series for production workers in manufacturing, summarized in the first two entries of table 2.1, is a BLS AWH series. Just as for AHE, AWH is available annually on an SIC basis until 2002 and on an NAICS basis from 1939 onward.

From 1932 onward, AWH embodies a paid-hours concept, and is part of what became known as the CES survey. The AWH series is described in Jones (1974, pp. 54–65) [in a comparison of the data-gathering procedures of BLS and COM, from the standpoint of "observed historical record on hours"], Greis (1984, pp. 279–281), Kirkland (2000, pp. 26–27), and Sundstrom (2006c, p. 2.307; 2006f,

p. 2.48; 2006g, p. 2.316). It is clear "how average weekly hours are calculated using CES data. For each industry [within manufacturing], the sum of the reported paid hours worked is divided by the total number of production workers reported for that same industry...When aggregating hours across industries, the average weekly hours for each industry are weighted (or multiplied) by the proportion of production or nonsupervisory workers in the industry division" (Kirkland, 2000, pp. 26–27). The paid-hours feature is unambiguous: "the payroll survey (CES)...measures hours paid for on each full- or part-time job, including hours paid for standby or reporting time, and equivalent hours for which the company paid vacation, holiday, or sick pay" (Devens, 1978, p. 4).

There are two defects with the BLS 1932-onward AWH series. First, by design, all paid hours are counted, whether or not work hours. Paid and work hours were essentially the same for most of U.S. labor history, but this concordance ended around the start of World War II: "The conceptual difference between earnings per hour at work and per hour paid for is important only after 1939" (Rees, 1960, p. 2). "Until the 1940s, the distinction in most industries between hours paid and hours actually worked was relatively unimportant. The widespread adoption of paid vacations of increasing length and of an increasing number of paid holidays (and in some industries paid travel time, lunchtime, etc.), however, has raised average weekly hours (which are hours paid for) above average hours worked by increasing amounts" (Sundstrom, 2006c, p. 2.307). The present study requires at-work rather than paid hours. Either the BLS AWH series must be converted from paid to at-work hours, or the series cannot be used.

Second, the existence of multiple-job workers under a survey of establishments (as is the CES) means that such employees are counted as multiple workers, one for each establishment in which a job is located. This point is made by Greis (1984, pp. 280–281) and Sundstrom (2006f, p. 2.48). The implication is an understatement of hours worked from the standpoint of individual workers, although a correct reporting from the standpoint of the individual establishments.

Jones (1963, p. 378) writes "the BLS historical series may be described as an 'implied' hours per week series before 1932, for which average weekly hours are equal to the ratio of average weekly earnings to estimates of average hourly earnings." This statement is correct literally only for the pre-1920 period; derivation of AWH is more complex for the other years.

Thus the BLS AWH figures for 1904, 1909, 1914, and 1919 are estimated indirectly, or implicitly, as the ratio of COM AWE to BLS

AHE. The computations are described in Bowden (1955a, pp. 802–804). Bowden does not state explicitly how AWE is obtained. He mentions "Census of Manufactures weekly earnings by industry" (1904), "weekly earnings for 1909, computed from Census data for all manufacturing," "average weekly earnings derived from Census data" (1914), and simply "weekly earnings in 1919." However, as the COM for these years produced only annual data, that, at best, could enable computation of average *annual* earnings (see *Regular Census* under EARNINGS AND WAGES), it must be that such AAE are calculated from the pertinent Census figures and then divided by 52 to obtain average *weekly* earnings. Computations of this nature are made in other contexts. Rees (1960, p. 8) works with "average weekly earnings derived from the full Census of Manufactures (average annual earnings divided by 52)." Jones (1961, p. 33) observes that "average weekly earnings would be equal to Census average annual earnings divided by 52."

Other things being equal, a direct computation is deemed to be of higher reliability than an indirect one. This judgment suggests that the 1904-to-1919 figures are of relatively low reliability. Another limitation is the differing nature of the numerator and denominator of the indirect computation. The numerator, based on Census data, is superior in quality to the denominator, which is BLS AHE.

It is logical to discuss AWH for 1923–1929 before AWH for 1920–1922 and 1930–1931, because the series for the latter periods are in part dependent on the series for the 1923–1929 period. The 1923–1929 BLS AWH is described in Bowden (1955a, pp. 805–806), Rees (1960, p. 16), Kendrick (1961, p. 44), and Jones (1963, p. 378). First, a combined series of AHE for 12 industries is generated, the individual-industry data based on BLS generally biennial surveys supplemented by interpolation or extrapolation and external information (especially Census figures and NICB series). Second, a combined series of AWE for the 12 series is obtained from BLS monthly payroll surveys (with Census benchmarks). Third, the ratio AWE/AHE yields AWH for the 12 industries (indirect computation again).

Fourth, for 1923 and 1929 (the only Census years during this period), using Census data, the ratio of average prevailing hours for all manufacturing to average prevailing hours for the 12 industries is taken. The Census does not provide average prevailing hours as such but only a frequency distribution of wage-earners by prevailing-weekly-hours intervals (see *Regular Census*); so the numerator and denominator of the ratio is computed by BLS. Fifth, for 1923 and

1929 again, total-manufacturing AWH is estimated as the product of the fourth-step ratio and the third-step 12-industry AWH. Sixth, for 1924–1928, total-manufacturing AWH is interpolated via the 12-industry AWH.

The technique is admirable for the logic of its procedure and its apparent effective use of available BLS and Census data. However, Rees (1960, p. 16) points out that a downward bias in the estimated AWH emanates from the third step in the procedure. The numerator, AWE, uses Census benchmarks and so has "reasonably full coverage of small establishments"; whereas the denominator, AHE, based essentially on BLS surveys, lacks such coverage and therefore is overestimated. Not mentioned by Rees, probably because he considers the point obvious, is that small establishments tend to have lower (weekly and hourly) earnings per worker than do larger establishments. Therefore the 12-industry estimated AWH is underestimated. This downward bias carries through all subsequent steps, so that total-manufacturing AWH is also underestimated.

Empirical evidence supports Rees' argument. First, Rees (1960, pp. 16–17) compares the BLS AWH series with two alternative series—his own (indirectly computed) and that of the NICB (see *Private Surveys* further ahead). For each year, 1923–1929, the BLS figure is the smallest. Second, Rees also shows the Census prevailing (full-time) hours for 1923 and 1929. Of course, these figures are higher than any of the other three, but they are unreasonably higher than the BLS: "It is hard to understand why in prosperous years firms would report 'prevailing hours' 12 or 13 per cent higher than average actual hours" (Rees, 1960, p. 17). Third, for the year 1929, Jones (1961, pp. 17–18, 116–127; 1963, p. 379) derives directly AWH for nine of the 12 industries underlying the BLS AWH. Using the BLS, indirect, technique, the nine-industry AWH is 44.3 hours; while direct computation raises AWH to 47.2 hours. Fourth, three external independent sources (NICB, Pennsylvania Department of Labor and Industry, National Safety Council) show AWH for 1929 uniformly above corresponding BLS figures (Jones, 1961, pp. 19–20; 1963, p. 379).

The BLS AWH series for 1920–1922 is discussed in Bowden (1955a, pp. 804–805) and again criticized in Rees (1960, pp. 18–19). The description is vague. BLS heavily relies on the hours data of King (1923) (see *Private Surveys* further ahead), but also makes use of Census prevailing-hours information for 1919 and 1923. Rees (1960, p. 18) interprets that "the BLS apparently did not rely wholly on King in obtaining the level of hours for 1920–22, but relied in part on its estimates for 1923–29. The result is that the BLS estimate is again the lowest available." Another problem is that the King series combines wage-earners and salaried workers. Unless corrected to include only wage-earners, the King figures probably overstate wage-earner AWH, as during a depression wage-earner hours presumably fall relative to salaried-worker hours. This problem partly counteracts the downward bias in AWH emanating from use of the 1923–1929 component of the BLS series.

For 1930–1931, as described in Bowden (1955a, p. 805), BLS AWH is obtained indirectly as the all-manufacturing AWE/AHE ratio. The numerator is obtained from BLS Census-adjusted payroll surveys; the denominator via interpolation between the all-manufacturing 1929 and 1932 figures, with the 12-industry AHE serving as the interpolator. One recalls that the 1929 AHE figure is too high, suggesting that the 1930–1931 AHE is also biased upward and, correspondingly, the 1930–1931 AWH biased downward.

National Compensation Survey (Employment Cost Index)

The BLS AHE series can be converted from a paid-hour to a work-hour basis via *division* by a work-hours/paid-hours ratio series. This use of such a ratio series is directly relevant to the present study, but, to the best of my knowledge, has not been advocated (at least in publication) either by BLS or by private researchers. In contrast, at least one academic scholar (Sundstrom, 2006g, p. 2.317) suggests that BLS AWH could be switched from paid hours to hours at work via *multiplication* by the work-hours/paid-hours ratio. In fact, since 1989, this technique has been utilized by the BLS in construction of its measures of factor productivity, thus adjusting and replacing the previously used paid-hours-based AWH in the measures (Jablonski, Kunze, and Otto [JKO], 1990, p. 17). All BLS data of the hours-worked/hours-paid ratio, for production workers in manufacturing, are summarized in table 2.3.

Beginning with data for 2001, the ECI component of the NCS is the source of the hours-worked/hours-paid ratio, for the purpose of the BLS productivity measures (BLS, 2003b), as suggested in the first entry in table 2.3. From the standpoint of the present study, there are several positive features of the ECI-based ratios. First, the ECI survey is well-established, so continuation of the ratio series is assured. Second, the quarterly frequency of the ECI survey enhances the representativeness of an annual series. Third, the quarterly results are available quickly—a desirable characteristic for an ongoing series.

Table 2.3 Ratio of hours at work to hours paid, 1947–2005, production and nonsupervisory employees,[a] manufacturing: BLS Data

Study	Period	Data Source
BLS, unpublished[b]	2001–2005[c]	NCS (ECI)
BLS (2001, Table 1)	1983–2000	HWS
BLS (1996, Table 1)[d]	1981–1995	"
JKO (1990, p. 22)[e]	1947–1988[f]	CPHM, SEEEC, HWS
BLS (1960, p. 11)	1958	CPHM
BLS (1969, p. 6)	1959, 1962, 1966	SEEEC
BLS (1971, p. 38)	1968	"
BLS (1973, p. 42)	1970	"
BLS (1975, p. 43)	1972	"
BLS (1977, p. 46)	1974	"
BLS (1980, p. 320)[g]	1976,[h] 1977	"

Notes:
[a] Also termed "production and related workers" and "nonoffice employees." Inclusion of nonproduction workers where stated.
[b] Data kindly provided by Mr. Shawn Sprague, of BLS.
[c] Figures for later years also presumably obtainable direct from BLS.
[d] Reprinted in Sundstrom (2006g, p. 2.316).
[e] Mixture of production and nonproduction workers in manufacturing. Reprinted, 1959–1980, in Sundstrom (2006g, p. 2.316).
[f] Interpolation and extrapolation for missing years.
[g] Figures can be approximated only, as ratio of pay for working time to sum of pay for working time, pay for leave time (except sick leave), and pay for sick leave—all figures dollars per hour, work-hours).
[h] Related to establishments employing 20 or more workers.
CPHM = Composition of Payroll Hours in Manufacturing.
HWS = Hours-at-Work Survey.
SEEEC = Survey of Employer Expenditures for Employee Compensation.

Negative characteristics of the ECI-based ratio also exist. First, "ratios from the ECI represent a combination of actual leave and leave practices" (BLS, 2003b, p. 2). Failure to incorporate fully changes in actual leaves lends a false stability to the ratio. Second, the ECI involves fixed weights. Therefore it "does not reflect changes in leave practices due to a shift in the mix of occupations, tenure and work schedules[,] which will usually lead to larger shifts in the hours at work ratio" (BLS, 2003b, p. 3). Presumably, eliminating the effect of shifts in employment also biases the measured ratio in the direction of greater stability. Third, the ECI sample is relatively small compared to the establishment (CES) sample that yields the AWH series. Fourth, although the ECI-based ratio has been published back to 1988, the figures are for durable and nondurable manufacturing separately, not for total manufacturing (BLS, 2003b, p. 6). Other positive and negative features are discussed in *Hours-at-Work Survey.*

Hours-at-Work Survey
Prior to the ECI-based hours-worked/hours-paid ratio, the BLS produced the Hours-at-Work Survey (HWS), which provided annual estimates of the hours-worked/hours-paid ratio for production and nonsupervisory workers in manufacturing. Initiated in 1982, the survey was terminated in 2000, and annual ratios were published for 1981–2000 (second and third entries in table 2.3). The HWS, based on a sample of establishments, is discussed in Kunze (1984, 1985), JKO (1990), BLS (1996, 2001, 2003b), and Sundstrom (2006g). Compared to the ECI, the HWS has the advantage of respecting the changing mix of establishments from period to period. Further, it is entirely retrospective, thus incorporating only actual leave, not leave "practice." BLS (2003b, p. 4) itself states that "the level of the hours-worked hours[-]paid ratio from the HWS is preferred."

Unfortunately, the HWS has conspicuous defects. First, although (like the ECI survey) it is *an* establishment survey, it is not part of *the* (CES) establishment survey. In particular, the sample size of the HWS survey was never more than two percent of that of the CES survey. (The ECI-based sample is significantly larger than that of the HWS, but nevertheless, still a small percentage of the CES sample.) Therefore, notwithstanding the fact that the HWS (like the NCS-ECI) is a stratified random sample, the sampling error of the hours-worked/hours-paid ratio (for both the HWS and ECI) is higher than that of AHE.

Second, the survey questionnaire and the technique of computer-assisted telephone interviews of nonrespondents to the survey were unsatisfactory, to the point that it was clear that reporting was not necessarily made on the stipulated basis of payroll records. Third, in part because of alterations to repair the survey, the response rate eventually deteriorated substantially. BLS (2001, p. 4) itself warned: "the continued decline in the HWS response rate from 75 percent in 1994 to 50 percent in 2000 should give users of these data reason for caution." It is not surprising that, as a consequence, the survey was terminated shortly thereafter, and replaced by the ECI, which presumably has a relatively high and stable response rate. Therefore, fourth, the HWS itself is limited to adjusting only twenty years (1981–2000) of the AHE. Nevertheless, with the ECI linked to the

HWS ratio, as done in BLS (2003b, p. 4), a continuous annual series from 1981 to the present exists in principle (and also in practice, given BLS cooperation—see notes b and c of table 2.3).

Employer Expenditures for Employee Compensation
As a by-product to its data on earnings and benefits, the surveys of Employer Expenditures for Employee Compensation (SEEEC) provide figures on the hours-worked/hours-paid ratio for nine years over the time span 1959–1977. A predecessor survey, Composition of Payroll Hours in Manufacturing (CPHM), supplies a figure for the year 1958. Together, they are the fifth and subsequent entries in table 2.3. As a group, these ratio figures can be viewed as a precursor to the HWS. There is a problem with the 1976 result, because small firms are omitted. More serious, considering the entire period 1947–1980, is the lack of continuous annual figures of the hours-worked/hours-paid ratio.

The second problem is addressed by JKO (1990, p. 22), who offer a continuous series of the hours-worked/hours-paid ratio for manufacturing for the 1947–1988 period, with the SEEEC used for 1959–1977 and the HWS covering 1981–1988. The JKO series (the fourth entry in table 2.3) is included as a BLS series, because BLS utilized their series in measures of productivity, as documented in BLS (1997, p. 91, note 10). AWH adjusted from a paid to at-work basis using the JKO series (followed by the HWS), for all workers in manufacturing, 1949–2003—but on a, superseded, SIC basis—is in BLS (2003a).

Unfortunately, the JKO series has deficiencies. The first, and a surprising, deficiency is the neglect of the CPHM figure for 1958—so that the JKO figure for that year misses being survey-based. In contrast, JKO are correct in not using the 1976 SEEEC figure, as it is based on only large establishments. A second deficiency is that the series is a mixture of production workers alone (for 1959, 1962, and the HWS years 1981–1988) and all workers (for other years). Fortunately, as observed in Sundstrom (2006g, p. 2.317), this problem is not serious: "where they overlap during the 1980s...the ratios for all employees and for production and nonsupervisory workers are very similar." A third limitation, definitely serious, is the extrapolation (at the two-digit SIC level) used to obtain the figures for 1947–1958. One would have hoped that the BLS had performed unpublished surveys of at-work hours for some of these early years, but apparently that is not the case. Sundstrom (2006g, p. 2.317) does not approve of the extrapolation and so reprints only the

1959–1980 segment of the JKO series: "Mary Jablonski, Kent Kunze, and their colleagues also extended both [total-nonfarm and manufacturing] series back to 1947 by assuming that the industry-specific ratios were constant prior to the date of the first survey—as there was no justification for assuming the ratios were constant, these earlier numbers have not been included in the series here." In contrast, the JKO linear interpolations for the intervening missing years (1960–1961, 1963–1965, 1967, 1969, 1971, 1973, 1975–1976, and 1978–1980) are quite defensible.

Fixed-Weight Hours Index and Subsequent Data
Corresponding to the BLS fixed-weight wage index for 1890–1907 is a fixed-weight weekly-hours index. The series is found in *Bulletin 77* (1908, p. 7), and discussed in Brissenden (1929, p. 353), Douglas (1930, pp. 73–76), and Kendrick (1961, p. 446). The fixed weights (proportional to industry payrolls in the 1900 Census) reduce the desirability of the series. The hours concept is full-time rather than actual: "The hours of labor given in this report represent the regular full time hours of the occupation—that is, the time that the employees as a class were engaged in work...it must not be inferred that all employees engaged in the establishments reported in this article worked full time" (*Bulletin 77*, 1908, p. 16). The description of Douglas (1930, p. 73) is "the normal number of hours constituting an accepted week's work."

Further on the positive side, the number of industries even in the early period (1890–1903), at 53, is pleasingly high. Also, for that early segment, industries of small size are included; so industry representativeness is assured.

The weekly-hours concept of BLS industry surveys for 1907–1918 is the same as that of the 1890–1907 survey—regular hours. Again, the Douglas (1930, pp. 81, 112, 113) description is "standard hours...the standard number of hours constituting a week's work...the normal working week in these industries."

First Annual Report of Commissioner of Labor
Again, we owe to Long (1960, pp. 36–38, 145–146) a convenient list of the average hours per day for each industry in the year 1885, in Commissioner of Labor (1886). Long also computes a (presumably employment-weighted) daily-hours average of all industries in the Report. Because of the large number of industries included in the Report, Long (1960, p. 37) is impressed with "the very comprehensive data of the First Annual Report," for this purpose.

Census

Regular Census

It is interesting that only for six regular Censuses did the Census Bureau publish data on weekly hours of wage-earners in manufacturing. These are the consecutive Censuses 1909, 1914, 1919, 1921, 1923, 1929. The data are in Bureau of the Census (1913, p. 306; 1917, p. 482; 1923, p. 70; 1924, p. 68; 1926, p. 1150; 1933, p. 51). Commentary on the data is in Brissenden (1929, pp. 351–356) and Kendrick (1961, pp. 442–445).

The explicit concept—satisfactory for this study—is full-time hours, what the Census itself calls prevailing or normal hours: "the number of hours normally worked by wage earners...the prevailing practice followed during the year in each establishment" (Bureau of the Census, 1913, p. 306; 1917, p. 482), "prevailing hours of labor per week" (1923, p. 69), "the number of hours normally worked each week, according to the prevailing practice during the year of the establishment reporting" (1926, p. 1150), "customary hours of labor" (1933, p. 51). Another advantage is that data are collected for all industries in the Census rather than for a subset or sample; and, as for other Census data, figures are presented not only for total manufacturing but also for the individual industries that compose manufacturing. Further, the response rate in terms of workers is high. For 1929 (the only year for which this statistic is provided), while 12 percent of establishments did not report their hours, these establishments constituted only one percent of employment. One can judge these Census data as encompassing a high degree of reliability.

There do exist problems with the Census hours data. The most obvious deficiency is that the data are not collected, and therefore not presented, as an average. As Brissenden (1929, p. 351) comments: "The Census Bureau does not show, as such, the average number of hours worked per week in different census years. It does, however, present material from which an average may be computed." The yearly data are shown in a frequency table. The number of wage-earners (average employment over the year) is distributed according to intervals of prevailing weekly hours of labor. All wage-earners in a given establishment are placed in the interval in which the establishment's prevailing hours fall—even though, of course, some workers may have experienced hours other than the norm. The Census Bureau views this variation of work-hours as unimportant: "In most establishments, however, the hours of labor are the same for practically all the wage earners" (Bureau of the Census, 1926, p. 1150).

More serious are some incongruous properties of the intervals of "prevailing hours of labor per week." First, the number of intervals is not quite uniform for all years. For the five censuses 1909–1923, the number is eight; for 1929, it is ten. Second, 1909, 1914 versus 1919, 1921, 1923 have different interval patterns. There is a certain logic to these changes, as the Bureau is recognizing that the weekly-hours distribution is moving to the left. Third, the lowest and highest intervals are open-ended; for example, "40 hours and under" and "over 54 hours," for 1929.

Private authors, making varying assumptions about mid-points of the hours intervals, use the Census all-manufacturing frequency distributions to compute AWH for various of the years. Results are in Brissenden (1929, p. 352), Wolman (1938, p. 2), Rees (1960, pp. 17–18; 1961, p. 36), and Kendrick (1961, p. 445).

Special Reports
Man-Hour Statistics for Selected Industries: The Census special reports on Man-Hour Statistics provide average monthly hours per wage-earner for all the "selected" industries combined and for each industry, for the years 1933, 1935, 1937, and 1939. Of course, the advantages and disadvantages of the earnings data in these reports are applicable here. Important is the actual-work hours concept, which is certainly preferred to the BLS paid-hours basis. An additional limitation is the, unusual, monthly frequency. Weekly hours were the customary, and superior, frequency at the time.

1880 Census: In the 1880 Census, manufacturing firms reported the "number of hours in the ordinary day of labor [from] May to November [and] November to May" (Wright, 1900, p. 315); but, except for the iron and steel industry, the Census did not report results. Fortunately, Atack, Bateman, and Margo [ABM] (2002), again preceded by Atack and Bateman (1992), use Census manuscripts to summarize the data see chapter 5, *Average Daily Hours*.

Atack and Bateman (1992, p. 132) make a strong case that the data pertain not to actual-work hours but rather to full-time (prevailing) hours: "we can be reasonably certain that firms did not report the actual hours at work on a particular day or over a period of time, since virtually all reported an integer number of hours. Instead, we interpret the question as asking for the normally scheduled hours of work..."

Weeks Report: As important a source as the Weeks Report is for nineteenth-century data on wages, it is also noteworthy for the inclusion of a specific section on "Hours of Labor" (Weeks, 1886, pp. xxviii–xxxiii). This section of the Report is discussed in Long (1960, p. 35) and Atack and Bateman (1992, p. 136). The hours concept is daily and full-time: "the number of hours constituting a regular day's work" (Weeks, 1886, p. xxix)—acceptable features for the present study. Prevailing hours are also the interpretation of Sundstrom (2006b): "the hours reported by Weeks are probably best thought of as the normal or scheduled hours of work, rather than actual hours worked or paid for."

Data are obtained for the period 1830–1880 for 53 industries (though arguably only 46 are manufacturing—see *Special Reports* under EARNINGS AND WAGES) and 37 states (plus the District of Columbia, DC), and results are shown in three separate tabulations: all-states (plus-DC) combined conjoined with all-industries combined, by individual state for all-industries combined, and by industry for all-states combined. The national all-industry summary is a pleasing improvement over the presentation of wage series.

Unfortunately, the information content is less impressive than that for wages. First, the data are not annual but only quinquennial: 1830, 1835,..., 1875, 1880, which simply reflects the questionnaire schedule (Weeks, 1886, p. xiii). Second, just as for the Census 1909–1929 data, averages are not shown. Rather, there is a frequency distribution of "statements" according to six daily-hours intervals. Means for the intervals are not provided; but, pleasingly, the upper and lower intervals (eight and less than nine hours per day, 13 and less than 14) are closed rather than open-ended. Nevertheless, any independent computation of an average (for a given year) is forced to incorporate arbitrary assumptions about the interval averages. For example, Long (1960, p. 35) takes the midpoint as the interval mean, while Sundstrom (2006b) (who, pleasingly, reproduces the Weeks national summary table) argues, convincingly, that a more-reliable mean is the lower bound of the interval: "Given the widespread tendency to set hours at whole numbers, however, a better average is probably obtained by taking the average of the lower bound of each interval again weighted by the number of statements." However, Sundstrom uses that technique to compute ADH only for 1830 and 1880.

A deficiency of the Weeks hourly data, as it is for his wage statistics, is the omission of employment figures. It follows that, in computation of the overall mean hours, the means of the hours intervals cannot be weighted by number of workers. Rather, all who use the Weeks frequency distribution weight by number of "statements." A given establishment could provide more than one statement, because the questionnaire schedule requests daily hours according to class of employees. Sundstrom (2006b) comments that "the most important classes of employe[e]s" (to quote the schedule; Weeks, 1886, p. xiii) reflect "presumably the most numerous occupations." That at least is logical, even though the number of workers in each occupation is not provided.

The number of statements increases uniformly over time, from 34 in 1830 to 1039 in 1880. That, in itself, suggests scope for an inconsistent quinquennial series. In principle, for a meaningful weighting pattern of hours intervals, one could (at least for later years of the time period) resort to the individual-industry figures and weight by Census-derived employment. However, again one would obtain biased and inconsistent results, because the number of industry-interval cells is so great relative to the number of statements. As one goes back in time, the number of empty cells increases dramatically. As one goes forward, the very increase of statements adversely affects the consistency of the constructed series.

Notwithstanding these problems, Weeks-based central-tendency daily hours for 1830 and 1880—at least as computed by Atack and Bateman (1992, p. 136)—are surprisingly close to the McLane Report figure for 1831 (see *Congress and Treasury* below) and identical to the Atack-Bateman Census figure for 1880. However, Atack and Bateman are inconsistent in taking a weighted average of hours-intervals midpoints for 1830 but the modal hours for 1880. A weighted average for 1880 would result in a Weeks figure close, but not identical, to their own.

Congress and Treasury

Aldrich Report: In the Aldrich Report (1893, pp. 178–179), Falkner presents a table of "average [daily] hours of labor" for each of 21 industries along with the all-industries unweighted average, for 1840–1891. With rare exception, the individual-industry series are continuous annually, although the beginning year varies. The Falkner (all-industries) series is discussed in Long (1960, p. 36), Whaples (1990, pp. 25–26), and Sundstrom (2006b; 2006f, p. 2.47).

Sundstrom observes that Falkner does not provide detail as to whether the hours reported are normal (scheduled) hours of work or actual hours worked or paid for. However, all the individual-industry series have runs of recurring figures and, for many of these series, these figures are whole numbers. These characteristics do suggest that the Falkner data pertain to full-time (scheduled or prevailing) hours, as do the Weeks data.

Long (1960, p. 36) is correct in stating that "the Aldrich group of firms is not necessarily representative of manufacturing in general." As mentioned in **Congress and Treasury** under EARNINGS AND WAGES, only 12 of the 21 industries are clearly within the manufacturing sector. Another issue, shared by the Weeks series, is the retrospective nature of the data. Falkner (in Aldrich Report, 1893, pp. 179–180) himself writes

> The reduction in the number of hours [from only 11.4 in 1840 to 10.0 in 1891, for all industries] seems hardly so considerable as might have been expected. It must be remembered that our figures refer to certain picked establishments, where, in view of the complete organization at an early date, it is possible that shorter hours made an earlier appearance than in the mass of work-shops. It may therefore be doubted whether these figures, absolutely correct as they are for the establishments in question, give a perfectly adequate picture of general conditions.

Whaples (1990, pp. 25–26) observes: "Both these [Weeks and Aldrich] sources [of nineteenth-century manufacturing hours], however, have been criticized as seriously flawed by problems such as sample selection bias and unrepresentative regional and industrial coverage. Another problem is that the two series differ in their estimates of the average length of the work week by as much as four hours." Therefore Sundstrom (2006f, p. 2.47) is right in warning that "there are many doubts about their representativeness."

Yet, on balance, all users of the Weeks and Aldrich hours data come to a sanguine conclusion. Long (1960, p. 36) writes that "the Aldrich and Weeks data may not greatly misrepresent the length of the workday." Whaples (1990, p. 26) notes that "there is general agreement that the Weeks and Aldrich series come close to reality in their broader implication that the length of the work week declined in virtually every decade from 1830 to 1900, and that the pace of this change was very erratic. Both show the 1850s to be the decade of the greatest reductions, both show that the length of the work week fell by about nine hours between 1830 and 1890." Sundstrom (2006f, p. 2.47) observes that "the Weeks and Aldrich data seem to be in line with other sources on hours from the period."

McLane Report: As noted both by Atack and Batman (1992, p. 136) and by Margo (2000a, p. 229), the McLane Report supplies the earliest data on hours of work in manufacturing, but without summary. Fortunately, Atack and Bateman provide an average figure (see chapter 5, *Average Daily Hours*). Because the McLane Report provides data on wages rather than earnings, it is reasonable to assume that the hours data pertain to prevailing rather than actual hours.

State Labor Bureaus

State labor bureaus developed hours-worked data along with their earnings series, but only from the 1870s (Whaples, 1990, p. 26; Atack and Bateman, 1992, p. 137). So, with national-level alternative series available by this time, the state data have little marginal value. As an exception, Jones (1963, p. 379) makes use of Pennsylvania hours information for the year 1929.

Private Surveys

National Industrial Conference Board: Corresponding to its AHE series, the NICB produced series of average weekly *actual-work* hours for production workers in manufacturing for 1914–1948, with the same industry composition and same missing observations. For a given component industry of manufacturing, the series is computed by dividing total hours worked by total number of wage earners. Just as for AHE, the all-manufacturing series is a weighted average of the separate industry series, with fixed employment weights based on the 1923 COM.

The series is tabulated in full in NICB (1950, pp. 340, 342), Bureau of the Census (1960, p. 94; 1975, p. 172), and Sundstrom (2006d). The series is discussed in Beney (1936, pp. 19–20), Jones (1963, p. 378), Bureau of the Census (1960, p. 82; 1975, p. 154), and Sundstrom (2006d).

For parts of the 1914–1948 period, the NICB also produced an average *full-time* work-week series for production workers in all-manufacturing. Beney (1936, p. 12) writes that "the nominal work-week shows the number of hours that are supposed to be worked under normal, full-time conditions"; but she does not present the series. The reason is that the series is published for 1914, 1920–1932 and 1937–1938 only. The explanation for the 1933–1936 hiatus: "During the depression...the full time hours reported were so low that the Board itself believed that some establishments were confusing full time with actual hours and for this reason discontinued the series from April 1932 to February 1937" (Wolman, 1938, p. 7). Apparently the final month for which the series was produced is January 1939 (see NICB, 1939: February, p. 33; March, p. 49).

Wolman is suspicious of the reliability of the series, in part for the reason stated above, in part because of the overrepresentation of large establishments and underrepresentation of the South. He refers to "a monthly series purporting to be full time hours, published by the National Industrial Conference Board, for 1914, 1920–1932, and now currently reported since February 1937" (Wolman, 1938, p. 10).

National Bureau of Economic Research: The King (1923) study provides estimates both of average actual hours per week and average full-time hours per week, for "all factories" for 1920–1921 (by quarters) and 1922 (first quarter). The series are in King (1923, pp. 82, 87). Average actual hours are obtained in the usual way, via payroll records, as the ratio of total number of employee-hours to number of employees. Of course, the same problems of King's AHE series exist here: the questionnaire technique, the limited time period, and the combining of production and nonproduction workers. In addition, King (1923, p. 81) reports that "many informants, however, confused this term [full-time hours] with 'hours actually worked' and, despite careful efforts to eliminate all errors, some probably remained among the data tabulated."

On the positive side, even though King (1923, p. 86) notes that "the records secured of hours actually worked per week are probably more accurate than are those of full-time hours," it is useful to have both actual and prevailing average hours data for the same sample of industries and establishments within each industry—apparently not a property of the corresponding NICB data. Rees (1960, p. 18) makes use of this feature of the King data.

National Safety Council: Jones (1961, pp. 20, 127–128; 1963, p. 379, esp. note 19) reports that the National Safety Council (NSC) began, "in the early 1920's," to publish data collected from manufacturing firms on employment and hours worked. Investigation of the present author suggests that these data were processed only for the years 1925–1930. Just as King does, the NSC combines both wage-earners and salaried workers. Another defect is the mixing of actual hours and full-time hours: "the instructions to firms by the National Safety Council permitted the reporting of scheduled hours of work if data on the hours actually worked were not available" (Jones, 1961, p. 127). In this respect, what for the King series occurs inadvertently happens by design for the NSC series.

Given its limitations and the existence of alternative data sources, the NSC data are of limited use. Jones (1961, p. 20) does judge that "the NSC data appear adequate for indicating the approximate level of hours in manufacturing in 1929."

Layer Study: Layer (1955, p. 43) offers a table of average hours worked per day in the cotton mills of Lowell, Massachusetts, for 1825–1915. Monthly averages are shown for various time periods. However, rather than an annual series, 17 periods are distinguished, only seven of which are individual years. It is unfortunate that Layer does not present his hourly data in the form of an annual series.

BENEFITS

Bureau of Labor Statistics

Although the ECI has a benefits component, that series is not discussed here. There is no need, because the benefits series under Employer Costs for Employee Compensation (ECEC) correspond to those under the ECI in availability and have the advantages of a current-weight rather than fixed-weight design and an absolute-cost rather than index-number denomination. Table 2.4 summarizes data sources for total-benefits series in manufacturing, with ECEC composing the first four entries.

Ideally, one would want the benefits data (and, in fact all data) to pertain to production workers (in manufacturing, of course); but, as table 2.4 shows, this restriction is not followed in the ECEC. For 1986–1987 and 2004–2006 (3Q), figures are available only for production and nonproduction workers combined. For discussion of the ECEC benefits series, one may consult Nathan (1987, pp. 4–7), Wiatrowski (1999, p. 33), and BLS (2000a, pp. 9–10; 2002, pp. 1–2).

The composition of ECEC total benefits is as follows (with the arrangement here different from that of BLS). There are three categories of company-based benefits: (1) insurance benefits (life, health, short-term disability, long-term disability), (2) retirement and saving plans (defined-benefit, defined-contribution), and (3) other benefits (collection of severance pay, supplemental insurance plans). Then there are (4) legally required benefits: social security and Medicare, unemployment insurance (federal, state), and workers' compensation. Also formally included in ECEC benefits are (5) paid leave (vacation, holiday, sick, other) and (6) supplemental pay (overtime and premium pay, shift differentials, nonproduction bonuses). In the present study, items (5) and (6) are classified under earnings, thus converting the ECEC "wages and salaries" component of "total compensation" from a regular-earnings to a gross-earnings concept (see *National Compensation Survey and Predecessor Series*).

Table 2.4 Total benefits[a] in manufacturing—Data sources

Series	Source	Years	Occupations	Denomination(s)
ECEC (NAICS basis)	BLS Web site[b]	2004[c]–	all	dollars per hour worked, percent of compensation
"	"	2006[c,d]–	production	"
ECEC (SIC basis)	"	1986–2003[e]	all	"
"	"	1988–2003[e]	blue collar	"
EEEC	BLS (1980, pp. 319–320)	1959, 1962, 1966, 1968, 1970, 1972, 1974, 1976,[f] 1977	production and related	dollars per hour worked, dollars per paid-hour, percent of compensation
employee benefits	CC[g]	1947–1977 (odd years), 1978–	all	dollars per paid ("payroll") hour, dollars per year per employee, percent of payroll
"	"	1987–1998	hourly-paid	"
national accounts	BEA Web site[h]	1929–	all (wage-earners and salaried workers)	percent of compensation[i]
compensation	ASM, Census Bureau Web site[j]	1967–1986, 2005–	all	percent of payroll[k]

Notes:
[a] Also termed "supplements" (to wages or wages-and-salaries).
[b] www.bls.gov.
[c] Quarterly only.
[d] Begins fourth quarter 2006.
[e] 2002–2003 quarterly only.
[f] Relates to establishments employing 20 or more workers.
[g] CC, Fringe Benefits, Employee Benefits, Employee Benefits Study; biennial 1947–1977, annual 1979–.
[h] www.bea.gov, National Income and Product Accounts, Tables 2.2A and 6.2A, B, C.
[i] Computable as ratio of "compensation of employees *minus* wage and salary disbursements" to "compensation of employees."
[j] www.census.gov.
[k] Computable as ratio of supplemental (fringe) benefits to payroll.
BEA = Bureau of Economic Analysis.
CC = U.S. Chamber of Commerce.

The concern here is with total benefits; so the ECEC occasional rearrangement of items within benefit components is irrelevant, for the total-benefits figure is unchanged. However, some changes appear to have been substantive. In 1998 "sickness and accident insurance" was renamed "short-term disability." That in itself does not matter; but "the definition was expanded to include all insured, self-insured, and State-mandated plans that provided benefits for each disability, including unfunded plans" (BLS, 2002, p. 2). Therefore this change widened the scope for, and therefore the amount of, total benefits. In 2006 the component "other benefits" was discontinued. This change obviously reduced the amount of total benefits. It is reasonable to assume that these changes had only a small effect on total benefits.

The measurement of benefits under ECEC is based on current costs and established usage: "the ECEC is a measure of the price of the benefit plans in March of each year multiplied by the usage (that is, the number of workers receiving the benefit) of the specific benefit plans that were established during the initial collection of data. (These initial data are used to reduce the amount of data a survey respondent must provide.)... If the mix of benefit plans changes, however, new usage patterns are obtained" (Wiatrowski, 1999, p. 33). "The annual cost is then divided by the annual hours worked to yield the cost per hour worked for each benefit" (BLS, 2000a, p. 10). To enhance data reliability, "a BLS economist does the initial collection of data through a personal visit" (Wiatrowski, 1999, p. 37, note 8) rather than having the employer respond to a mailed questionnaire.

Neither ECEC nor the other data sources shown in table 2.4 directly provides a series of the mark-up of benefits over earnings, that is, the benefits/earnings ratio. However, given that all sources publish earnings and benefits as complementary series, the benefits/earnings ratio series is readily computed. A serious problem with the benefits/earnings ratio—and indeed with the definition of AHC as the sum of AHE and AHB (see chapter 1, EARNINGS VERSUS BENEFITS)—is the assumption that employer cost of benefits is equal to employee valuation of the benefits. As Rees (1959, p. 13) states: "Including wage supplements along with wages in figuring worker compensation implicitly assumes that the benefits provided by both public and private insurance and welfare plans are worth what they cost."

All the available data sources of benefits provide only employer cost, as the very phrase "employer costs" in ECEC (and "employer expenditures" in EEEC) imply. The fact that one has no choice in the matter does not mean that one cavalierly adopts the assumption "employer-cost equals employee-value." Famulari and Manser (1989, p. 28) examine the issue and reach a balanced conclusion: "we conclude that employer cost is limited as a measure of employee value. For some purposes, however, using employer cost to proxy the median worker's value of non-legally required benefits seems to be a reasonable approximation to employee value.... Employer cost as a proxy for how the median employee's value of benefits has changed over time also seems reasonable."

Benefits under the Employer Expenditures for Employee Compensation (EEEC) survey are shown in the fifth entry in table 2.4. Discussion and summary presentation of EEEC benefits are in BLS (1976a, pp. 176–177; 1980, pp. 319–320), Nathan (1987, pp. 6–7, 11, note 5), and Wiatrowski (1999, p. 33). With current-employment weights, absolute-dollar denomination, and separate data for production (and related) workers, EEEC benefits are naturally complementary to the ECEC benefits, although the breakdown of EEEC benefits differs from that of ECEC. With an inherent gross-earnings rubric, EEEC benefits are correctly defined for the present study.

The components of EEEC benefits are: (1) retirement programs (social security, private pension plans), (2) health-benefit programs (life, accident, and health insurance; workers' compensation; other [principally state temporary disability insurance, not shown separately]), (3) unemployment-benefit programs (unemployment insurance, supplemental unemployment-benefit funds), (4) savings and thrift plans, (5) vacation and holiday funds. Item (5) is a very small, at most one-tenth of one percent of compensation; it is different from payment for leave time itself (this payment included in gross earnings).

"The EEEC and the ECEC are different surveys that use different methods" (Wiatrowski, 1999, p. 33). This remark is particularly true regarding measurement of benefits. Unlike ECEC, EEEC considers the employer cost of a benefit to be the actual expenditure on the benefit for the year under consideration. This, retrospective, approach provides a true count of cost; but it requires precise expenditure data. The ECEC technique, based on price and usage, possesses the advantage of being current, ongoing, and forward-looking. The EEEC, unlike the ECEC, survey was conducted entirely by questionnaire. Finally, the EEEC survey, again unlike the ECEC, is available only for disconnected years (see table 2.4), and has been discontinued.

U.S. Chamber of Commerce

The U.S. Chamber of Commerce (CC), a private organization, has assembled and published benefits information since 1947. The data of this group are summarized in the sixth and seventh entries of table 2.4. Benefits are broadly defined, including even "break time," and the benefits categories have changed over the years (wherefore no attempt is made to list them here). The expenditure method of recording benefit payments is used. In addition to the pertinent publications of the CC itself, sources for discussion of the CC benefits series are Rees (1960, pp. 23–25) and Greis (1984, pp. 283–284).

The advantage of the CC series is its availability for years in which the EEEC and ECEC surveys were not taken (although, until 1978, the CC provides data only for alternate years). However, the CC figures have disadvantages that mitigate against their use. First, except for 1987–1998, data for production ("hourly paid") and nonproduction ("salaried") workers are commingled, although the salaried employees are those for whom the firm reports compensation on an hourly basis. Second, the sample is small—for example, only 191 reporting manufacturing firms in 2005. For the hourly paid workers, a subsample of the overall sample is employed. Third, the sample is not stratified by size or other characteristic. Fourth, reliance is entirely on voluntary compliance via questionnaire. Fifth, as Rees (1960, pp. 24–25) argues, the sample overrepresents large firms. In its early history, the CC sample excluded firms below a stipulated number of employees (500, in 1947–1953; then 100), although by 2005 this restriction had been removed. Importantly, Rees provides empirical evidence that the rate of response is higher for larger-size firms and that such firms make larger payments for benefits.

Bureau of Economic Analysis

The national accounts, produced by the Bureau of Economic Analysis (BEA), provide aggregate data that enable computation of total benefits as a percent of total compensation, the next-to-last entry in table 2.4. Total benefits (termed "supplements to wages and salaries") in manufacturing are obtainable as "compensation of employees" (available in Tables 6.2A, B, C, line 13) *minus* "wage and salary disbursements" (available in Table 2.2A, B, line 4). Division by "compensation of employees" (and multiplication by 100) yields total benefits as a percent of compensation, comparable to the "percent of compensation" denomination for other entries in table 2.4. Others— Rees (1960, pp. 21–23), Bureau of the Census (1960, p. 84), Jones (1961, pp. 38, 39, esp. note 2; 1963, p. 385), Moehrle (2001, p. 14)— have used or discussed the national-accounts information whether implicit in the above aggregate series or explicit in the disaggregate series mentioned below.

Advantages of the national-accounts data are the methodical collection of information (data permitting), the inherently appropriate attention to firms of all sizes, and, for aggregate percent of compensation, a continuous annual series from 1929 onward. Another positive feature is the strict adherence to a gross-earnings concept, with the consequent narrow definition of benefits.

The main deficiency is the merging of wages and salaries, and the merging of "supplements to wages" and "supplements to salaries." There are other limitations. First, there is only a limited disaggregation of supplements to wages and salaries for manufacturing: "employer contributions for social insurance" and "employer contributions for employee pension and insurance funds" (Tables 6.10, B, C, D, and 6.11, B, C, D; BEA Web site). Components of the two disaggregate flows are not provided for manufacturing. Second, even these two flows are shown only from 1948 onward. For disaggregation in earlier years, one must resort to data for all industries combined, available 1929–1938 as follows (with exclusion of component items inapplicable to manufacturing): (1) employer contributions for social insurance (with disaggregation into components: old-age and survivors insurance, state unemployment insurance, federal unemployment tax, cash sickness compensation funds) and (2) other labor income (with disaggregation into components: compensation for injuries, employer contributions to private pension and welfare funds) (U.S. Department of Commerce, 1954, p. 210).

Census

Like the BEA series, Census data on benefits (the final entry in table 2.4) are at the aggregate level. However, unlike the BEA series, the Census data are not continuous. They are provided only for two discrete time periods, 1967–1986 and 2005 onward. It is also unfortunate that, just as for the BEA series, the Census figures are not reported separately for production workers—perhaps surprisingly, because the wages component of total compensation is so reported. Fringe benefits include employer costs of almost all conceivable benefits. Exclusions are specified, as "such items as company-operated cafeterias, in-plant medical services, free parking lots, discounts on employee purchases, and uniforms and work clothing for employees" (American FactFinder Help, undated-a).

CHAPTER 3

Existing Earnings and Wage Series

The data sources, mainly official, for earnings and wage series are presented in chapter 2, EARNINGS AND WAGES. To the extent that formal earnings or wage series were developed by the data-gathering institutions themselves from their own collected data, these series are also discussed in EARNINGS AND WAGES. Private scholars have used these data to construct their own earnings and wage series, which are arranged here in three sections with associated tables: average hourly earnings (AHE) or average hourly wage (AHW) series, average daily wage (ADW) series, and composite series.

AVERAGE HOURLY EARNINGS OR HOURLY WAGE RATE

Rubinow: Table 3.1 summarizes private hourly wage or earnings series that are uniform in nature (meaning not a composite of other series). Isaac M. Rubinow (1914), the first entry, develops a series to replace and extend the *Bulletin 77*, 1890–1907 series (see chapter 2, *Wages*). The Rubinow series is discussed in Rubinow (1914, pp. 803–808), Douglas and Lamberson (1921, pp. 409–410), and Rees (1961, pp. 8–9). Running from 1890 to 1912, the Rubinow series is an unweighted average of 15 individual-industry series, each of which is continuous over 1890–1912. Even though the *Bulletin 77* series is based on many more industries and is a weighted index, nevertheless the correspondence between the *Bulletin 77* and Rubinow series (for 1890–1907, of course) is remarkably close. The maximum divergence is only slightly over one percent (computed from table in Rubinow, 1914, p. 809). Therefore Rubinow (1914, p. 808) judges his series to be "a proper measure of the fluctuations of wages in American manufacturing industry in general."

Table 3.1 Average hourly earnings or hourly wage rate,[a] production workers[b] in manufacturing: Private uniform series,[c] 1890–1935

Study	Series	Period	Denomination	Data Source	Weights[d]
Rubinow (1914, p. 809)	AHW	1890–1912	index number (1890–1899 = 100)	BLS	unweighted
Douglas and Lamberson (1921, p. 415)	AHW	1890–1918	"	"	"
Brissenden (1929, pp. 123, 356)	AHE	1899, 1904, 1909, 1914, 1919, 1921	dollars per hour[e]	COM, BLS	employment
Douglas (1930, pp. 96, 101, 108)[f]	AHE; industries: all, payroll, union,	1890–1926	"	BLS	"
Creamer (1950, pp. 44–45)	AHWRI	1922–1935[g]	index number (April 1923 = 100)	"	1922–1923: number of establishments 1923–1935: employment

Notes:
[a] Excludes benefits or supplements.
[b] Also termed "wage-earners" or "manual workers."
[c] Annual, except where otherwise noted. Direct computation of hourly figures, except where otherwise noted.
[d] Current, except where otherwise stated.
[e] Indirect computation of hourly figures: ratio of average weekly earnings to average weekly hours, see text.
[f] Reprinted (all, payroll, and union industries) in Bureau of the Census (1960, p. 91; 1975, p. 168). Reprinted (all industries only) in Bureau of the Census (1949, p. 67) and Margo (2006g). Reprinted (all and payroll industries), 1914–1919, in Rees (1960, p. 19).
[g] Monthly.
AHWRI = average hourly wage-rate index.
BLS = Bureau of Labor Statistics.
COM = Census of Manufactures.

There are some problems with the Rubinow series. First, one of the industries (building trades) is outside manufacturing. Second, the series is inherently inconsistent, combining the "union" hourly wage rates of six industries with the "payroll" hourly wages of nine others. Third, for the "union" industries, furthermore, data are payroll-based wages for 1890–1907, union wage rates only for 1908–1912. Fourth, the individual-industry series are computed non-uniformly from component (occupational) series. Some series are employment-weighted averages, some unweighted averages, of occupational wage or earnings series. Fifth, as an index number, the Rubinow series is not as informative as if it were denominated in absolute terms (dollars per hour).

Douglas and Lamberson: Paul H. Douglas and Frances Lamberson (1921), the second entry in table 3.1, modify the Rubinow series and extend it six additional years, to 1918. The new series is discussed in Douglas and Lamberson (1921, pp. 410–415), Hansen (1925, p. 27, note 2), and Rees (1979, pp. 917–918). Douglas and Lamberson find that earnings or wage data in 1912–1918 are published by Bureau of Labor Statistics (BLS) for only ten of Rubinow's 15 component industries, and the base period is the year 1913 rather than 1890–1899. For these industries, the unweighted average of occupational wage series is taken, and the resulting individual-industry series is linked to the corresponding Rubinow series via the overlap for 1912, thus converting the base period from 1913 to the Rubinow 1890–1899. The Douglas-Lamberson series, for all years 1890–1918, is the unweighted average of wages or earnings for the ten industries. The authors show that the difference between their 10-industry and Rubinow's 15-industry series is small, concluding that "the results are substantially the same whether one uses ten industries or fifteen (Douglas and Lamberson, 1921, p. 412).

The Douglas-Lamberson series has the same deficiencies as the Rubinow series. In particular, there are seven "union" industries and three "payroll," one industry having switched from payroll to union status. An additional limitation is that three industries, all "payroll," lack observations for 1915 and 1917. It should be noted that the "union" versus "payroll" nomenclature to identify an industry is first applied in Douglas (1930).

The main virtue of the Douglas Lamberson series is historical; it represents the great Paul Douglas' "first substantial empirical work on wages" (Rees, 1979, p. 917). Another praiseworthy characteristic is the computation of AHE, for the payroll industries, as the ratio of total earnings to total number of hours worked (instead of taking the hourly wage rate per se); the result is a gross-earnings concept. Also, Hansen (1925, p. 27, note 2) accepts the "general reliability" of the Douglas-Lamberson series, on the basis of comparison with the *Bulletin 77* series and an index of male wages (from various sources) in King (1923, pp. 203–204).

Brissenden: The third entry in table 3.1 is the Paul F. Brissenden AHE figures, a unique series the construction of which is described in Brissenden (1929, pp. 282, 292–297, 354, 356). The series is discussed in Achinstein (1930, p. 371), Douglas (1930, pp. 593–598), Bowden (1955b, p. 920), and Rees (1961, p. 24, note 9). Brissenden (1929, p. 356) derives AHE indirectly as the ratio of average full-time weekly earnings to average prevailing (full-time) hours worked per week. The denominator, full-time weekly hours, is obtained in a straightforward fashion. Census data are used to compute prevailing hours for 1909, 1914, 1919, 1921 (see chapter 2, *Regular Census* under DAILY, WEEKLY, OR MONTHLY HOURS OF WORK) and the Rubinow (1914, p. 810) full-time hours series is taken for 1899 and 1904. The Rubinow series is linked to the Census-based figures via the 1909 overlap. (Both Rubinow and Douglas-Lamberson provide index numbers of full-time weekly hours, the series corresponding in every way to their respective wage series shown in table 3.1.)

The numerator, average full-time weekly earnings, is not so readily understood. Its construction is described here in stepwise fashion.

Step 1: Brissenden (1929, p. 282) computes Census unadjusted average annual earnings (AAE) in manufacturing for 1899, 1904, 1909, 1914, 1919, 1921. The resulting figures are the same as in Douglas (1930, p. 219). The AAE figures are converted to relative form (1904 = 100) (Brissenden, 1929, p. 297).

Step 2: Average weekly earnings (AWE) under full operation in 1904 are obtained from *Bulletin 93* (1908, p. 11)—see Brissenden, 1929, p. 297)—the amount is $10.06 (see chapter 2, *Special Reports* under EARNINGS AND WAGES).

Step 3: The month in which the highest number of manufacturing wage-earners were employed in 1904 is found to be October (Brissenden, 1929, pp. 294–295; from Bureau of the Census, 1907, p. lxxix). The proportion of trade unionists in New York State employed in that month is 0.892, which is one minus the unemployment proportion, the latter figure from the New York Bureau of Labor Statistics (Brissenden, 1929, p. 294).

Step 4: From occupational data in the 1890 and 1900 Censuses, the proportion of wage-earners in "manufacturing and mechanical industries" unemployed during some part of the year in New York and in the entire United States is obtained. Taking complements yields the proportion of wage-earners employed. The ratio of the U.S. employment proportion to the New York employment proportion is 0.977 and 0.973 for 1890 and 1900, with the average 0.975 (Brissenden, 1929, pp. 294, 296).

Step 5: The product 0.892 x 0.975 = 0.870 is the estimated ratio of *actual-to-full* employment for the United States in October 1904, the month of highest employment during that year. Therefore the estimated ratio of *full-to-actual* employment is the inverse of 0.870, and full-time weekly earnings are estimated as $10.06/0.870 = $11.56 (Brissenden, p. 300).

Step 6: For the other Census years, average full-time weekly earnings are computed as the product of relative AAE (from step 1) and $11.56 (Brissenden, 1929, p. 297).

Brissenden's procedure is complex and unnecessarily so. There is no rationale to develop figures on a *full-time* basis for weekly earnings and weekly hours, given that *actual* AHE is estimated as the quotient. As Rees (1961, p. 24, note 9) comments, "Brissenden's method [of estimating hourly earnings] is extremely and needlessly complicated." Even if a rationale is assumed, there are problems with the procedure. In step 3, the New York trade-unionist unemployment rate is unreliable in itself and unrepresentative nationally, as both Douglas (1930, p. 597) and Rees (1961, p. 24, note 9) assert. In step 4, unemployment during the year is overstated, because unemployment is counted irrespective of duration. Thus U.S. unemployment is computed as 21.0 and 27.2 percent, and New York unemployment 19.1 and 25.2 percent, in 1890 and 1900. To justify the procedure, one must assume that the percentage overstatement is approximately the same in New York and the entire country.

In step 5, it is implicitly assumed that "if only 87 per cent were employed, it also represents 87 per cent of a full-time week of those on the payroll" (Achinstein, 1930, p. 371). The conjectured one-to-one correspondence between percent of workers employed and percent of full-time hours worked by employees on payroll is at variance with the empirical evidence. Also, one recalls that the weekly earnings of $10.06, emanating from *Bulletin 93*, pertain to "a 'composite' week, made up of the 'busiest' weeks during the year in the reporting establishments... And yet Brissenden assumed that, even during this composite 'busiest' week, the time actually worked was only 87 percent of full time" (Bowden, 1955b, p. 920).

Finally, step 6 involves the implicit assumption that "the census average wage is considered to be a good measure of the changes in full-time weekly or yearly earnings" (Achinstein, 1930, p. 371). Achinstein shows that Brissenden's (1929, p. 280) claim that "the Census average wage...does appear to reflect the changes in...full-time earnings" is based on Brissenden's own computations that *assume* the relationship in question.

Douglas: The AHE work of Paul H. Douglas (1930, pp. 73–110) is the fourth entry in table 3.1. The Douglas series are discussed and criticized in Latimer (1930, p. 483), Wolman (1932, pp. 400–403), Bowden (1955a, p. 803; 1955b, pp. 918–919), Bureau of the Census (1960, p. 79; 1975, p. 151), Rees (1961, pp. 11, 19–23, 60–62; 1979, p. 919), and Margo (2006g). The Douglas AHE series in table 3.1 break new ground in several respects, even though their wage and earnings ingredients emanate entirely from BLS publications. First, the series pertain to manufacturing strictly defined, with no extraneous trades or industries. Second, overall manufacturing is divided explicitly into "union" industries and "payroll" industries, depending on whether worker compensation is represented by union wage rates or payroll-based hourly earnings (although payroll data, from *Bulletin 77*, are used for union industries 1890–1906). Separate series are constructed for "union" and "payroll" manufacturing (composed of six and eight industries, respectively) as well as for overall manufacturing. Third, individual-industry series are employment-weighted averages of absolute-level (not relative, index-number) occupational wages or earnings, and the aggregate union, payroll, and overall manufacturing series are obtained as employment-weighted averages of the individual-industry series. Fourth, the series fill a gap that exists even in current BLS average series, namely, a continuous manufacturing series for 1908–1919.

The Douglas series are not without problems. Consider first payroll industries. First, earnings data have missing years, and interpolation via AWE or AAE is employed. Second, until 1914, data are limited, generally to industry-specific occupations, and exclude most unskilled workers. The solution is linking via the overlap in 1914. Third, the iron-and-steel industry throughout excludes segments that happen to have lower average earnings (see Rees, 1961, pp. 60–62). Fourth, establishment coverage and occupational categories for the clothing industry varies over time (Latimer, 1930, p. 483). Fifth, the sample is small in early years and biased toward large establishments.

Turning to union industries, the main problem is that union wage rates are both higher and more stable than actual earnings of all workers in the union industries. Indeed, this deficiency applies also to actual earnings of unionized workers alone. Another defect is the overweighting of union industries in the overall manufacturing series. Instead of union membership, the total number of skilled and semi-skilled workers serves as weights. The overweighting is about 200 percent (see Wolman, 1932, p. 400).

Because of the above criticisms, the Douglas aggregate series have been viewed generally as unreliable. A typical judgment is that of Bowden (1955a, p. 803): "The Douglas estimates of hourly earnings are too high as to level and their trend is biased downward over the period (1890 to 1926)." However, later work has resurrected the aggregate *payroll* series. A prescient statement is: "it may sometimes be desirable to use only... [average hourly earnings in 'payroll' industries] to represent all manufacturing" (Bureau of the Census, 1960, p. 79). Empirically, Rees (1961, p. 37) compares the Douglas payroll series to his own all-manufacturing AHE series for 1890–1914, and concludes that there is "a similarity between our series for all manufacturing and Douglas's series for payroll industries that is astonishing in view of the very different sources and methods used."

Creamer: Based on the BLS recording of percentage changes in wages rates for 1919–1935 (see chapter 2, *Wages*), Daniel Creamer (1950, pp. 5–6, 40–43) constructs three indexes of wage rates in manufacturing, shown as the last entry in table 3.1. The indexes are discussed in Rees (1960, p. 15, note 5). Creamer converts the wage-rate changes to indexes of wage-rate levels. Within establishments, wage changes are weighted by number of workers. At higher aggregation, Creamer's weighting patterns are delimited by the BLS recording. The indexes are: 1919–1923, 13 combined industries, wage changes weighted by number of firms; 1922–1923, all manufacturing, wage changes weighted by number of firms; 1923–1935, all manufacturing, wage-changes weighted by number of workers. Creamer assumes that establishments not reporting a wage change made no such change, and he does not link his indexes. The fact that the indexes pertain strictly to wage rates rather than to earnings further detract from their use in the present study.

AVERAGE DAILY WAGE RATE

Falkner: Prior to 1890, compensation series pertain almost entirely to wage-rates rather than earnings and to frequencies lower than hourly. Table 3.2 presents ADW series for the 1860–1890 period. The first two entries represent the work of Ronald P. Falkner, an academic who served as statistician for the Aldrich Report. He generates two overall wage-rate series: an unweighted average of series for 21 industries and an average of 17 industries weighted by Census data: employment

Table 3.2 Average daily wage rate,[a] production workers[b] in manufacturing: existing series, 1860–1890[c]

Study	Data Source	Period	Denomination	Weights
Falkner (Aldrich Report, 1893, pp. 174, 176)[d,e]	Aldrich Report (1893)	1840–1891	index number (1860 = 100)	unweighted
Falkner (Aldrich, Report, 1893, p. 176)[d,f]	"	"	"	gainfully employed (national level)
Mitchell (1908, pp. 170, 204–206)[e,g]	"	1860–1880[h]	index number (1860 = 100)[i]	unweighted
Mitchell (1908, p. 170)[e,g,j]	"	"	"	employment (reporting establishments)
Mitchell (1908, p. 177)[k]	Weeks (1886)	1860–1880	index number (1860 = 100)[l]	unweighted
Phelps Brown and Hopkins (1950, p. 277)[m]	Aldrich Report (1893)	1860–1891	index number (1860 = 100)	fixed weight; employment (firm level), gainfully employed (national level)
Long (1960, pp. 121–124)[n]	Aldrich Report (1893)	1860–1890[h]	dollars per day	employment (firm and state level), gainfully occupied (national level)
Long (1960, pp. 129–130)	Weeks (1886)	1860–1880	"	unweighted average (firm and state level), gainfully occupied (national level)
Long (1960, pp. 135–136)	Bulletin 18 (1898)	1870–1890	"	gainfully employed (national level)

Notes:
[a] Excludes benefits or supplements.
[b] Termed "manual workers" or "manual labor."
[c] Annual, except where otherwise stated.
[d] Reprinted 1860–1880 in Mitchell (1908, p. 170), 1860–1891 in Bureau of the Census (1960, p. 90), and (1860, 1865, 1870, 1875, 1880) in Long (1960, p. 20).
[e] Of 21 component industries, 6 (building trades, city public works, dry goods [clerks in stores], groceries, railroads, sidewalks) are clearly nonmanufacturing, and 1 industry (gingham) is included within another (cotton). Also, 2 industries (lumber, stone) are partly or entirely nonmanufacturing—see text.
[f] Of 17 component industries, 4 (building trades, city public works, dry goods [clerks in stores], railroads) are clearly nonmanufacturing, and 2 (lumber, stone) are partly or entirely nonmanufacturing—see text.
[g] Reprinted (1860, 1865, 1870, 1875, 1880) in Long (1960, p. 20).
[h] January and July.
[i] Central tendency measured not only as mean but also, alternatively, as median.
[j] Reprinted, mean series only, in Bureau of the Census (1960, p. 90).
[k] Reprinted (1860, 1865, 1870, 1875, 1880) in Long (1960, p. 25).
[l] Central tendency measured as median.
[m] Reprinted (1860, 1865,…, 1890) in Long (1960, p. 20).
[n] Of 13 component industries, 2 (lumber and stone) are partly or entirely nonmanufacturing—see text.

(1840, 1850), persons attached to ("engaged in") the industry (1860, 1870, 1880), with interpolation and extrapolation for other years. Descriptions and criticisms of the Falkner series are in Aldrich Report (1893, pp. 172–177), Bowley (1895, pp. 369–376), Abbott (1905, pp. 340, 343–350), Mitchell (1908, pp. 92, 104, 169–174), Phelps Brown and Hopkins (1950, p. 267), Bureau of the Census (1960, p. 79), Long (1960, pp. 17–21), Lebergott (1964, pp. 290–292), and Douty (1984, p. 18).

All the limitations of the Aldrich data (see chapter 2, *Congress and Treasury* under EARNINGS AND WAGES) are present in Falkner's series—and there are further problems. Several of Falkner's industries are clearly outside manufacturing. Restricting oneself to the 17-industry series, "building trades," "city public works," "dry goods (clerks in stores)," and "railroads" are obviously nonmanufacturing. Further, Lebergott (1964, p. 291, notes 8, 10) argues in effect and convincingly that two other industries are incorrectly placed in manufacturing. "'Stone'... is really a mining industry if we may judge from its occupational component (quarrymen, laborers)... one of the two firms constituting the... lumber manufacturing industry (the New York firm) was actually a distributor." There are also defects in the weighting pattern. Interpolation is crude; extrapolation could not take advantage of the 1890 Census; and "white lead," used primarily in paints, is weighted by persons attached to all chemical manufacturing. In addition, Falkner takes only January, sometimes only July, data—rather than averaging for the two months.

The most widespread criticism of Falkner's series is equal weighting of occupations in the computation of individual-industry series. Lebergott (1964, p. 290) writes, in somewhat of an exaggeration: "his procedure ended up in an extreme overweighting for the rates paid to the limited group of overseers and foremen, for whom a phenomenal number of wage quotations were secured, and a massive underweighting of wages for semiskilled and unskilled employees. Since the wage trends for the two groups differed, the result was a biased wage index."

The Lebergott criticism should be tempered. First, the importance of the resulting bias would differ between industries. Second, the extent of the bias would vary depending on the time interval considered. For example, for 1860–1865, Long (1960, p. 20) finds a difference of eight percent between Falkner's 17-industry series and his own carefully constructed Aldrich-based series. However, if one considers 1860–1861 and the Long January figures, then the difference is trivial—only 7/10ths of 1 percent. The result shows that, at least around the end of the antebellum period, the Falkner figures possess a certain reliability. An advantage of the Falkner series is its existence for the antebellum period (although the number of industries and establishments decreases as one goes further into the past).

Mitchell: Wesley C. Mitchell (third and fourth entries in table 3.2) improves the Falkner 21-industry series by using the Aldrich reported-employment data to weight occupations within an industry. There are four series: unweighted and weighted, each with mean and median as alternative measure of central tendency. Mitchell (fifth entry) also creates a Weeks-based 30-industry series, with the median. Mitchell's series are discussed in Mitchell (1908, pp. 92–121, 175–191), Phelps Brown and Hopkins (1950, p. 267), Bureau of the Census (1960, p. 79), Long (1960, pp. 18–25), and Lebergott (1964, pp. 289–290). Mitchell's Aldrich-based series are praised by Lebergott (1964, p. 289) as "a now-neglected work of indefatigable endeavor and broad economic insight." However, the Mitchell series are not without problems. First, nonmanufacturing series are included: for the Aldrich-based series, all that are in Falkner's 21 industries; for the Weeks-based series, "iron mining" and "ship-carpentry." Second, the median is rarely used to develop wage series, and for good reasons. "The median is laborious to compute and, though free of the randomness that derives from the influence of extreme values, is subject to the randomness that derives from gaps in the distribution of the values in a smaller number of observations, . . . [and] wage data of manual workers are free of really extreme values" (Long, 1960, p. 19). Third, Mitchell errs in using Aldrich-reported rather than Census employment to weight industries: "after all, the purpose of the average should be to give most weight to industries that employ the most workers in the nation as a whole and not to the ones that happened, through accident of selection and reporting, to be the largest employers in the Aldrich sample" (Long, 1960, p. 21). Fourth, the Weeks-based individual-industry series, for lack of reported employment data, of necessity involves equal weighting of occupations for each establishment.

Phelps Brown and Hopkins: The Ernest H. Phelps Brown and Sheila V. Hopkins series (sixth entry in table 3.2) is a variant of the Mitchell Aldrich-mean series, for a longer time period. It differs from the Mitchell series in three respects: industry composition (14 industries, of which building trades, railroads, and stone are nonmanufacturing); persons-engaged rather than Aldrich-reported employment to weight industries; fixed weights (1870–1879 for occupations, 1870 for industries) rather than current weights. The last characteristic effectively makes the series non-usable in the present study. For discussion, one can consult Phelps Brown and Hopkins (1950, pp. 267–270) and Long (1960, pp. 18–19).

Long: In an impressive synthesis and extension of manufacturing wage series for the 1860–1890 period, Clarence D. Long (1960, pp. 15–25, 32) develops series respectively based on Aldrich, Weeks, and *Bulletin 18* data (last three entries in table 3.2). The series are critiqued in Lebergott (1961b, pp. 263–264; 1964, pp. 291–297, 303–304) and Williamson (1975, pp. 18–22).

For the Aldrich- and Weeks-based series, Long restricts his input data to occupations, industries, and firms available for the entire 1860–1890 or 1860–1880 periods. For Aldrich, this means 13 industries and 49 firms; for Weeks, 18 industries and 67 firms. Consider a given industry. For each establishment, the employment-weighted (Aldrich) or unweighted (Weeks) average of occupational wages is computed. For each state, the employment-weighted (Aldrich) or unweighted (Weeks) average of firms' wages is computed. To construct the national wage, each state wage is weighted via the Census number of gainfully employed in the industry in the state (linearly interpolating between Censuses). Finally, the overall manufacturing wage is obtained by again weighting the industry wages by the number of gainfully employed nationally in the respective industries.

Several aspects of the Long series warrant praise: denomination of wages in dollars instead of index numbers, strict definition of manufacturing, and use of Census weights. Impressive is the stated and faithfully followed methodology of using data characterized by "continuousness over a long period for the same nominal occupations, firms, and industry" (Long, 1960, p. 12).

However, the series are susceptible to strong criticism. First, the number of industries and of firms within some industries is small compared not just to the Census but even to the Aldrich and Weeks totals. Even so, two of the Aldrich industries, stone and lumber, are better classified as nonmanufacturing (see Falkner). Second—and an especially focused criticism of Lebergott—is the overweighting of skilled workers compared to unskilled workers in both the Aldrich-based and Weeks-based series. Third, the geographical limitations of the Aldrich and Weeks data are inherent in Long's series. Fourth, Lebergott (1964, p. 295) argues that, for the 1860–1870 decade, both logic and empirical evidence show that "the census data are clearly to be preferred to the inadequate and biased Aldrich [per Long] reports." Logically, "one does not customarily prefer a sample consisting of one establishment to a sample of 12,000 establishments, nor a sample covering 15 percent of all employees to one with 100 percent of all employees" (Lebergott, 1964, p. 292). Empirically, "where the Aldrich samples are small...for every industry but one [and, more generally, for 9 out of 11 of the 13 Long industries]...the Aldrich materials point to much greater gains [in wages] than do the census figures" (Lebergott, 1964, p. 303—see also p. 292).

Yet these criticisms should be tempered. First, "Long uses only a selection of manufacturing industries to represent all manufacturing. However, he [Long, 1960, pp. 32–35] demonstrates carefully that the resulting difference is unimportant" (Lebergott, 1964, p. 291, note 8). Second, Williamson (1975, p. 18) shows that "the Aldrich-Long (adjusted) series...behavior...is consistent with what we know about the determinants of occupational wage structures." Third, external data can be used to correct the Northeast restriction of the Aldrich-based series (see chapter 5, *Interpolator and Extrapolator Series*). Fourth, unfavorable comparison with the Census does not exclude use of the Long-Aldrich series to *interpolate between Census years* (again see chapter 5, *Interpolator and Extrapolator Series*). In fact, it is arguable that the restriction of the Long sample to a fixed set of firms enhances the reliability of that specific use.

To construct the series based on *Bulletin 18*, Long (1960, p. 17) selects ten occupations which "might be called manufacturing occupations" and weights each individual occupation-city series by Census gainfully employed in the state in which the city is located (presumably again with intercensal interpolation, though he does not state this). The reliability of the *Bulletin 18* data is controversial (see chapter 2, *Wages*). Nevertheless, clearly, Long makes effective use of these data to construct a manufacturing wage series.

Composite Series

It is impossible to generate a truly long-run wage or earnings series without resort to more than one data source and method of construction. The resulting series segments would be linked, thereby constructing a composite series. The series of this type for AHE (or AHW) of production workers in manufacturing are shown in table 3.3.

Rees: For his all-manufacturing AHE, Albert Rees (1960, 1961) uses a technique that is unique for a composite series for any economic variable. He constructs a uniform series for 1890–1919 and then a composite series for 1932–1957. The series are linked via interpolation over the intervening years (1920–1931). The resulting full composite series is shown as the first entry in table 3.3. The 1890–1919 segment is presented and discussed in Rees (1960, pp. 19–20; 1961, pp. 23–40), Douglas (1962, pp. 446–447), Lebergott (1961a, p. 774), Jones (1963, pp. 380–381), Bureau of the Census (1975, pp. 154–155), Allen (1992, pp. 126–127), Hanes (1992, p. 275), Margo (2006a), and Sundstrom (2006a). Construction of the 1890–1919 segment is best understood as a stepwise sequence.

Step 1: Considering the Census years 1889, 1899, 1904, 1909, 1914, and 1919, Rees adopts the definition of manufacturing in the 1939 Census of Manufactures (COM). Therefore he makes deductions from the respective Census manufacturing totals of (1) annual earnings of wage-earners and (2) average number of wage-earners. First, figures pertaining to nonmanufacturing industries (with one minor exception) are deleted for each Census. Second, the 1889 Census includes hand-and/or-custom-trade establishments in manufacturing, and the associated figures must be removed. Pure hand and/or custom trades are excluded as nonmanufacturing. For industries partly hand and/or custom trade and partly factory, removal of the hand-and/or-custom-trade component is conducted via the 1899 industry-by-industry trade/factory wage-payments and employment proportions. Logging establishments, excluded from the Census lumber industry in 1889, are reinstated via the 1899 lumber-industry including-versus-excluding logging wage-payments and employment proportions.

Step 2: For 1909, 1914, and 1919, (2) average number of wage-earners in seasonal industries is corrected for undercounting due to zero Census figures for months in which the plant is not in operation. This is because the Rees methodology requires average employment on a time-of-plant-operation rather than full-year basis (shown in step 6 below).

Step 3: For the six Census years, the quotient (1)/(2), defined in step 1, is AAE, where (1) and (2) are corrected according to steps 1–2.

Step 4: AAE in intercensal years is estimated via an interpolator series based on data from labor bureaus of three states: Massachusetts (1889–1919), Pennsylvania (1892–1919), and New Jersey (1895–1919), with (unstated but presumably) weighting by employment and (stated) appropriate linking in 1892 and 1895. Prior to computation

Table 3.3 Average hourly earnings or hourly wage rate,[a] production workers[b] in manufacturing: Composite series (dollars per hour)

Study[c]	Series	Period	Component Series	
			Period	Source
Rees (1960, pp. 3–4, 19; 1961, p. 33)[d]	AHE (all manufacturing)	1890–1957	1890–1919	AAE[e] divided by product of ADO[f] and average full-time hours per day[g]
			1920	extrapolated via Creamer series
			1921–1931	interpolated via NICB series
			1932–1939	BLS AHE series adjusted via data of Census Man-Hour Statistics
			1939–1947	adjusted BLS weekly earnings[h] divided by adjusted BLS weekly hours[i]
			1947–1957	ASM and COM, total wages divided by total hours worked. Figure for 1948 interpolated via BLS AHE series.
Rees (1961, p. 72)	AHE (combined industry)[j]	1890–1914	1890–1914	state data,[k] method same as for 1890–1919, industries weighted by employment
Rees (1960, pp. 3–4)	AHE	1914–1957	1914, 1919–1957	BLS AHE series
			1915–1918	Douglas (1930) all-manufacturing series as interpolator
Hanes (1992, pp. 276–277)	AHW, AHE	"	1865–1890	Long (1960)[l]
			1890–1907	Douglas (1930)[m]
Hanes (1992, pp. 276–277)[n]	AHW[o]	1865–1907	1865–1890	Aldrich Report (1893)
			1890–1907	Bureau of Labor and BLS reports

Continued

Table 3.3 Continued

Study[f]	Series	Period	Component Series	
			Period	Source
Hanes (1996, pp. 856–857)[p]	AHE[o]	1923–1990	1923–1935	NICB (Beney, 1936)
			1935–1990	BLS "Hours and Earnings" and "Employment, Hours, and Earnings"

Notes:

[a] Excludes benefits or supplements.
[b] Also termed "wage-earners."
[c] Excluded are the composite series of Hansen (1925), which intermixes component series of hourly, daily, and weekly frequency, and of Phelps Brown and Hopkins (1950), which covers the entire economy rather than just manufacturing. With reason, Bowden (1955b, p. 919) states that "little can be said in favor of the Hansen index." The manufacturing segment of the Phelps Brown and Hopkins series is shown in table 3.2.
[d] Reprinted, 1890–1914, in Bureau of the Census (1975, p. 172) and Margo (2006a, p. 2.268). Preliminary series in Rees (1959, pp. 15–16).
[e] Years 1889, 1899, 1904, 1909, 1914, 1919 from COM. Computed as ratio of total wage payments to average number of wage-earners. Census data adjusted by deducting figures for nonmanufacturing industries and hand and/or custom trades. Other years via interpolation using data for three states (Massachusetts, New Jersey, Pennsylvania).
[f] Same state data source as for interpolation of AAE.
[g] Years 1909, 1914, 1919 from COM. Other years: interpolation or extrapolation using BLS data and data in Wolman (1938) and Douglas (1930).
[h] Adjusted to COM average weekly earnings.
[i] Converted to hours worked, using COM and NICB data.
[j] Nine-industry series. Serves as check on 1890–1914 segment of composite series.
[k] Connecticut, Maine, Massachusetts, New Jersey, Ohio, Pennsylvania, Rhode Island, South Carolina, Wisconsin.
[l] Aldrich-based daily wage-rate series, July figures (see table 3.2) divided by average hours per day (Long, 1960, p. 37).
[m] All manufacturing industries; incorrectly referenced to "union" industries in Hanes (1992, p. 281).
[n] Reprinted in Hanes (1993, p. 753) and Margo (2006j).
[o] Fixed weights, employment in 1889.
[p] Reprinted in Margo (2006j).

ADO = average number of days of operation.
NICB = National Industrial Conference Board.

of the series, nonmanufacturing industries are removed from the state total-manufacturing earnings and employment figures. Thus there is a continuous AAE series for 1889–1919.

Step 5: For each of Massachusetts, Pennsylvania, and New Jersey—annually for 1890–1919, 1892–1919, 1895–1919, again using state-labor-bureau data—Rees computes average number of days of operation (ADO) as the employment-weighted average of individual-industry figures (an improvement over the establishment-weighted averages provided by some state labor bureaus). To combine the state series for a continuous annual 1890–1919 ADO series, Census employment weights are used in Census years, linear interpolations of these weights in other years. Presumably, this same combination technique is used in step 4. Also presumably, there is linking here as in step 4.

Step 6: Average daily earnings (ADE), 1890–1919, is the ratio of AAE (from step 4) to ADO (from step 5).

Step 7: Rees develops an average-daily-hours (ADH) series, termed a "general hours series," for 1890–1919 with several component parts. Because daily earnings pertain to days in operation, consistency allows that hours be full-time or prevailing hours rather than actual work hours (but see chapter 5, *Average Daily Hours*)—and the data Rees uses are of that nature. These data are all weekly hours; Rees converts them to daily hours by division by six. Thus the "general hours series" is average full-time hours per day.

For 1909, 1914, 1919, Rees computes average hours from the Census frequency distributions of weekly hours (see chapter 2, *Regular Census* under DAILY, WEEKLY, OR MONTHLY HOURS OF WORK), after having removed nonmanufacturing industries according to the 1939 definition of manufacturing. For 1915–1918, interpolation is performed using the Douglas (1930, p. 114) series on weekly hours in (eight) "payroll" industries. For 1903–1914, an interpolating series is based on BLS data for seven industries, six of which are for payroll industries as processed by Douglas. The individual-industry series are combined by Census employment weights, with linear interpolation for intervening years. The resulting interpolator series is adjusted to pass through the 1909 and 1914 values of the general-hours series. For 1890–1902, Wolman's (1938, p. 2) 48-industry employment-weighted BLS-data-based series is linked to the general-hours series via the 1903 overlap.

Step 8: AHE, 1890–1919, is the ratio of ADE (from step 6) to ADH (from step 7).

The AHE 1932–1957 composite series has three components: 1932–1939, 1939–1947, 1947–1957. Considering 1932–1939, for each year of Census Man-Hour Statistics (1933, 1935, 1937, 1939—see chapter 2, *Special Reports* under EARNINGS AND WAGES), Census/BLS individual-industry AHE ratios are averaged using employment weights. The average ratio is applied to the BLS all-manufacturing AHE (see chapter 2, *Current Employment Statistics Survey: Average Hourly Earnings*). The 1933 ratio is used for 1932; the average of adjacent Census years for 1934, 1936, and 1938. Note that this technique converts BLS AHE from a paid-hour to work-hour basis.

The 1939–1947 segment is obtained via an involved procedure. First, the hours-worked/hours-paid ratio is estimated. For 1939, the 1932–1939 methodology is applied to Census and BLS average weekly hours (AWH). For 1943 and 1946, the ratio is estimated via information in NICB publications. The (1940–1942, 1944–1945) ratios are linearly interpolated between (1939 and 1943, 1943 and 1946). For 1947, the ratio is computed directly from Census and BLS all-manufacturing AWH. Second, for 1939–1947, BLS AWH is adjusted to a work-hour basis by multiplication by this ratio. (For 1947, the result is Census AWH itself.) Third, the all-manufacturing Census/BLS AWE ratio is computed for 1939 and 1947 and linearly interpolated for 1940–1946. (Presumably the Census 1947 AWE figure is obtained by dividing AAE by 52.) Fourth, for 1939–1947, BLS AWE is multiplied by this ratio to obtain adjusted AWE. Fifth, AHE on a work-hour basis is the ratio of adjusted AWE to adjusted AWH.

For 1947–1957, AHE on a work-hour basis is derived as production-worker total wages divided by total hours, per the Annual Survey of Manufactures and COM (see chapter 2, *Regular Census* and *Special Reports*, both under EARNINGS AND WAGES). The 1948 figure is estimated using BLS AHE as the interpolator.

To complete the full composite AHE series, the 1890–1919 and 1932–1957 segments must be joined. Using the Creamer series (see AVERAGE HOURLY EARNINGS OR HOURLY WAGE RATE), Rees extrapolates the 1890–1919 segment to 1920. He then adopts the NICB AHE series (see chapter 2, *Private Surveys* under EARNINGS AND WAGES) as an interpolator; but the NICB series is too high. It is adjusted downward to composite 1920 and 1932 figures (linking factors 6.4 and 11.4 percent). The adjustment factors for 1921–1931 are obtained via linear interpolation.

Rees deserves great credit for several characteristics of his work. Adoption of a composite-series framework permits him to generate a manufacturing production-worker AHE series for a longer time span than had ever been done. Basing the series on Census data provides a strong foundation of reliability. For 1890–1919, the methodology of moving from Census AAE to AHE "is an ingenious system" (Lebergott, 1961a, p. 774). The strict and consistent definition of manufacturing and the rigorous adherence to work hours rather than

Table 3.4 Average annual earnings, interpolation of intercensal years, wage-earners in manufacturing

Study	Period	Coverage of Interpolator Series		Listing of Series (pages in study)	
		No. States	Percent of Wage-Earners[a]	Interpolator	Final
Douglas (1930)	1889–1914	4–8[b]	16, 25, 29, 30, 30[c]	223	230[d]
	1915	6[e]	—	not listed	239[d]
	1916–1919	1[f] for 22 industries, entire U.S. for 13 industries[g]	48[h]	"	"
Rees (1961)	1889–1914	3[i]	7, 16, 16, 21, 26[j]	32[k]	33[l]
Rees (1960)	1915–1919	"	unstated	not listed	19

Notes:
[a] In Census years. Census years are 1889, 1899, 1904, 1909, 1914, 1919. Computation made by present author, based on data in Douglas (1930, pp. 219, 225, 235) and Rees (1961, p. 32).
[b] Connecticut (1889–1907), Iowa (1895–1914), Massachusetts (1889–1914), New Jersey (1892–1914), Ohio (1892–1908), Pennsylvania (1892–1914), South Carolina (1909–1914), Wisconsin (1889–1908).
[c] Numerator, denominator in Douglas (1930, pp. 225, 219). Hand and/or custom trades included (1889).
[d] Reprinted, 1890–1919, in Bureau of the Census (1949, p. 68; 1960, pp. 91–92; 1975, p. 168) and Margo (2006b, p. 2.271).
[e] Iowa, Massachusetts, New York, Pennsylvania, South Carolina, Wisconsin.
[f] New York.
[g] BLS data.
[h] Numerator is U.S. employment in all 35 industries; therefore coverage is overstated. Numerator, denominator in Douglas (1930, pp. 235, 219).
[i] Massachusetts (1889–1919), New Jersey (1895–1919), Pennsylvania (1892–1919). Nonmanufacturing industries excluded.
[j] Numerator and denominator in Rees (1961, p. 32). Logging included in 1889.
[k] Census years only.
[l] Reprinted in Bureau of the Census (1975, p. 172) and Margo (2006a, p. 2.268).

paid hours are praiseworthy. Estimation of average hourly benefits for 1929–1957 (see chapter 6, 1929–2006) is yet another admirable accomplishment.

Still, limitations of Rees' series cannot be ignored. The main problem with the 1890–1919 segment is the use of data from only three states to interpolate Census AAE. For 1890–1914, Douglas uses data from eight states for this interpolation. Rees (1961, pp. 31–32, note 16) justifies his limitation on grounds of consistency with state ADO data; but there is an opportunity cost in geographic and worker coverage, as shown in table 3.4. In particular, one observes that Rees' states are all in the Northeast, while half of Douglas' states are outside that region. For the 1916–1919 period, Douglas makes use of BLS national data, adding greatly to coverage. Related to this issue is Rees' failure to utilize—although he does obtain and discuss—Census national ADO data for the year 1904.

Allen (1992, p. 127) argues that the 1890–1919 segment of the Rees AHE series has a pro-cyclical bias, because workers may be reported on the payroll even when not at work. However, he does not demonstrate this bias for the Rees series itself. Allen's empirical evidence, supported by Hanes (1992, p. 275), pertains to a replication of Rees' methodology in 1972–1987. Also, Lebergott's (1961a, p. 774) (mild) criticism that Rees fails to make use of the Dewey Report to establish benchmark weekly earnings for 1890 ignores the serious deficiencies of that report (see chapter 2, *Special Reports* under EARNINGS AND WAGES).

In contrast, the 1932–1957 segment of the composite series is beyond criticism, with its effective use of Census and BLS data sources. Furthermore, the "linking" from the 1932–1939 to the 1939–1947 component and the "linking" from the 1939–1947 to the 1947–1957 component are seamless. In other words, hypothetical linking ratios in 1939 and 1947 are unity by construction. No such happy construction or consistent methodology applies to the joining of the 1890–1919 and 1932–1957 segments. As Allen (1992, pp. 126–127, note 3) comments: "the observed rate of change in wages in the Rees series after 1919 is actually the sum of the actual rate of change in the underlying data and the rate of change induced by Rees' splicing and interpolation procedures." Also, the use of the Creamer and NICB series requires accepting their principal deficiencies: the wage-rate rather than earnings aspect of Creamer; the underrepresentation of small firms and the South, as well as the fixed-weight characteristic of NICB.

For 1890–1914, Rees acknowledges and possibly alleviates the criticisms by constructing two alternative series to check on his composite series. A combined-industry series (second entry in table 3.3) results from applying the methodology to individual industries and then combining the industry series via Census employment weights that are linearly interpolated for intervening years. The combined series incorporates nine industries (with "textiles" consisting of seven sub-industries). Data are from nine states, but most provide "usable data only for some industries or only for short periods of time" (Rees, 1961, p. 40). ADH for the industries is obtained from a variety of sources: Census data, BLS data, Douglas series, and the "general hours series."

The combined-industry series is discussed in Rees (1961, pp. 40–73, 131–147), Margo (2006a, p. 2.269), and Sundstrom (2006a, p. 2.303). The main limitation of the series is stated by Rees himself: there are no data for leading industrial states (such as New York, Michigan, and Illinois) nor for important industries (such as automobiles, clothing, and meat-packing). Yet the closeness of the composite and combined-industry series is remarkable: "the two series never differ by more than one cent" (Rees, 1961, p. 71). In contrast, for 1914–1957, the composite series is decidedly different from the "BLS AHE series with the Douglas all-manufacturing series interpolator for missing years" (third entry in table 3.3).

<u>Hanes:</u> Three composite series constructed by Christopher Hanes constitute the final three entries of table 3.3. The series are presented in Hanes (1992, pp. 269–270, 275–281, 289–291; 1993, pp. 734, 752; 1996, p. 845) and Margo (2006i, 2006j). The Long-Douglas series (third-of-last entry), for 1865–1907, simply links the Douglas all-manufacturing AHE series (though Hanes mistakenly references the Douglas union-industries AHE) to the Long 13-industry Aldrich-based July series (though Hanes describes the linking as being in the opposite direction) via the 1890 overlap. The Long-Douglas series is provided purely as a comparison series to the Aldrich-BLS series (second-of-last entry).

The Aldrich-BLS series is obtained by matching Aldrich to Bureau of Labor and BLS wage-rate data for 27 occupations in seven industries (though Hanes says six). Each BLS occupational series is linked to the Aldrich July series, again via the 1890 overlap. The resulting 1865–1907 occupational series are averaged with equal weights to derive the individual-industry series. Then the industry series are employment-weighted via 1890 Census data.

The NICB-BLS series (last entry) is also constructed in two steps. For 21 industries, NICB AHE is linked to the BLS equivalent via the January 1935 overlap. Then the 21 industries are weighted by (presumably BLS) employment in 1989. Thus both the Aldrich-BLS and NICB-BLS series are of a fixed-weight nature—deliberately so for Hanes' analyses, but of doubtful use in the present study.

CHAPTER 4

Plan for Construction of New Series

This chapter presents the plan for construction of the average hourly compensation (AHC) series and a general outline of the construction procedure. First, the general characteristics of the series are discussed. Some properties of the series are determined from the methodological considerations discussed in chapter 1. Other properties have an operational nature. Second, the fundamental steps in generating the series are shown sequentially. Only the general outline of the procedure is provided. For justification of the procedure adopted and for full details of data construction, chapters five and six should be consulted.

METHODOLOGICAL PROPERTIES

Methodological ingredients of the series emanate from chapter one. The series is to be annual, continuous, and ongoing. Data are to be specific to manufacturing, to the maximum extent that data permit. Where manufacturing-industry wages are unavailable for the purpose at hand, recourse may be had to occupational data—but then both skilled workers and unskilled workers must be represented. Data should also be specific to production workers, but manufacturing industry is the priority (better all-workers in manufacturing than production-workers in a broader sectoral context).

Earnings are preferred to wage rates, and gross earnings to regular earnings (wherever the choice is available). Current weights are highly favored over fixed weights, and the absolute level of earnings or wages (dollars or cents per hour) over an index number. The hours concept is decidedly actual-work hours rather than paid hours. However, for the pre-1920 segment, the construction technique suggests acceptance of full-time work-hours. Where only regional (in practice, Northeast) manufacturing wages are available, other regional wage information is used to convert wages from the Northeast to a national geographic basis. The series is composite: uniformity is sacrificed for length.

Operational Characteristics

1. The generation of average hourly earnings (AHE) and the generation of average hourly benefits (AHB) are performed separately, so that the use of available information can be optimized—the best-possible data are used for each component of AHC.
2. First, the AHE series is constructed; then the benefits/earnings ratio (the mark-up of benefits over earnings) is computed. The product of that ratio and AIIE yields AHB.
3. An ongoing series implies the adoption of official data for the latest time segment. In this respect, regular-Census production-worker data are utilized to the maximum extent. Why is the Census chosen over BLS as the official source? The Census not only (in principle) is a complete count rather than a sample of manufacturing establishments but also fits the methodological principles more closely. Other Census studies are used as appropriate.
4. The methodology of Albert Rees (1960; 1961), both for earnings and benefits, is the default foundation of the new series. Any departure from the Rees series and technique requires justification.

Sequential Outline

Average Hourly Earnings

<u>1920–2006:</u> The period from 1920 onward is distinguished by computation of AHE without the use of benchmark figures or interpolator series. For 1947 onward, the Rees methodology is followed. AHE is the ratio of Census total earnings to total work-hours. For 1932–1946, the Rees AHE is replicated. The first divergence from Rees occurs for the 1920–1931 period, for which AHE is calculated as the ratio of BLS average weekly earnings to the average weekly work-hours series of Ethel B. Jones.

<u>1800–1919:</u> For the pre-1920 period, Rees is followed in the generation of data for benchmark (Census) years, with new interpolation of intervening years. Rees (1961, p. 24) writes, and the present author approves, "we accept the annual earnings estimates, and reject the kinds of hourly earnings estimates that have been built up from occupational data. One reason for doing this is that the coverage of the annual earnings data and the data on days in operation is very

much broader." However, the present study differs from Rees in utilizing annual-average daily or monthly earnings (or wage) series to interpolate intercensal figures. Also, in contrast to studies such as Long (1960) and Margo (2000b), such earnings or wage series are of interest not for their own sake but rather for interpolation and extrapolation purposes.

Benchmark years span 1820–1919. The goal is to generate series segments internally consistent and linked to the adjacent later segment via a one-year overlap. The segments are 1889–1919, 1849–1889, and 1820–1849. Benchmarks are Census years, except for 1831.

The procedure begins with Census benchmark figures for average annual earnings (AAE) in manufacturing. In this respect, it is similar to Rees. In fact, all the Rees AAE figures are taken (for 1889, 1899, 1904, 1909, 1914, and 1919). (The Rees AHE series begins in 1890, but the AAE component is shown one year earlier.) For 1849, 1859, 1869, 1879, and 1889 (the last for purpose of linking), AAE is specially calculated for the present study. AAE for 1849–1919 is for all workers and for the entire country. For 1820, 1831, and 1849 (the last, again for linking purpose), AAE pertains only to adult males and to the Northeast; these data are constructed by Kenneth L. Sokoloff and Georgia C. Villaflor. To repeat, 1831 is the only year for which the figure is obtained from a non-Census source, the McLane Report.

For the same benchmark years, over 1820–1919, the average number of days of operation (ADO) of manufacturing is obtained—via adjustment of the figures computed by Rees (1899–1919) and Jeremy Atack, Fred Bateman, and Robert A. Margo [ABM] (1869, 1879), via estimation specifically for this study (1849, 1859, 1889), and directly from Sokoloff and Villaflor (1820, 1831, 1849).

For all benchmark years, average full-time daily earnings (ADE) are the ratio AAE/ADO. (Earnings are full-time, because they are per day of plant operation rather than per day of the full calendar year.) For 1820, 1831, and 1849 (again, the last for purpose of linking), ADE for adult males in the Northeast is converted to an all-worker basis utilizing the studies of Claudia Goldin, Kenneth L. Sokoloff, and Georgia Villaflor. Then the Northeast all-worker ADE is adjusted to national coverage via the regional wage series of Robert A. Margo.

To complete the benchmark figures, a composite average full-time daily hours (ADH) series is developed especially for this study. The series is needed not only for the benchmark figures but also (below) for interpolator series. Therefore ADH is constructed annually for 1800–1919. The series has the foundation of its own benchmark figures: 1831 (from Atack and Bateman), 1879–1880 (from ABM), and 1890 (from Rees). Interpolator series are Aldrich-based (1840–1890) and Weeks-based (1830–1840), with simple extrapolation (1800–1830). So, for ADE benchmark years 1820–1919, AHE is computed as ADE/ADH.

Interpolator and extrapolator series are needed to extend benchmark figures to other years. Two approaches for constructing AHE for the 1890–1919 period exist in the literature. First, there is direct estimation. The only series encompassing the entire time period are the three (union, payroll, all) manufacturing series of Paul Douglas. Second, there is the indirect-estimation technique of Rees, whereby AHE is constructed as AAE/(ADO·ADH) with interpolation of AAE and ADO between benchmark years and separate generation of ADH. The present study adopts a compromise technique, employing Rees' methodology only for benchmark years and using the Douglas payroll series (extended back to 1889) for the purpose of interpolation. The Douglas series also serves to link the 1800–1919 and 1920–2006 segments of AHE.

For 1859–1889, the interpolator series is developed especially for the study. The 1860–1890 Aldrich-based series of Clarence Long is revised, estimated back to 1859, converted from daily to hourly via ADH, and adjusted from the Northeast to the full United States via regional wage series of Philip R. P. Coelho and James F. Shepherd (1859–1880) and of *Bulletin 18* (1881–1890).

For 1820–1859, the interpolator series is constructed from the Brandywine wage data of Donald R. Adams Jr. and the iron-industry wage data of Jeffrey F. Zabler, both converted from a monthly to an hourly basis. These sources are also used for extrapolation to the year 1800. Coverage is extended from male workers to all workers via relative-wage data and other information of Goldin and Sokoloff (1808–1859) and from the Northeast to the entire country via Margo's regional wage series (1820–1859).

Average Hourly Benefits

1929–2006: AHB is constructed as the product of the benefits/earnings mark-up and AHE. Sources of the mark-up ratio (except for the Rees series) are shown in table 2.4. For 1986–2006, the source is Employer Costs for Employee Compensation (ECEC); but two adjustments are in order. First, net wages and total benefits are converted to gross wages and residual benefits. Second, for 1986–1987 and 2004–2006, ECEC data pertain only to all manufacturing workers; the all-manufacturing ratio is converted to the production-worker ratio via suitable overlap.

For 1929–1957, the mark-up ratio emanates from the Rees series, itself based on Bureau of Economic Analysis (BEA) and Bureau of Labor Statistics (BLS) information. For 1959, 1962, 1966, 1968, 1970, 1972, 1974, and 1977, the ratio is computed from Employer Expenditures for Employee Compensation (EEEC) figures. Missing years between 1957 and 1977 are interpolated using the BEA national-accounts, all-worker manufacturing, mark-up over wages and salaries. Finally, figures for 1978–1985 are estimated via linear interpolation between 1977 and 1986.

1900–1928: The two components of pre-1929 benefits are "workers'-compensation benefits" and "pension-and-welfare benefits." For each component, an index is constructed relative to 1929. The annual estimate of the benefits amount is the product of the index and the amount in 1929. For workers' compensation, the index is the amount of insurance premiums. Data are from the studies of Price V. Fishback and Shawn Everett Kantor and of Arthur H. Reede. For pension-and-welfare benefits, the index is the percent of workers covered by pension plans in manufacturing. The index is based on data of Samuel H. Williamson.

CHAPTER 5

Average Hourly Earnings

1920–2006

Average hourly earnings (AHE) denote earnings on a gross basis and exclusive of benefits. For 1947 and 1949–2006, the Rees methodology is followed (see chapter 3, COMPOSITE SERIES). AHE is computed directly as the ratio of total wages to total hours, for production workers in all manufacturing, Census data. The computed series, denominated in dollars per hour, is inherently current-weighted, which is desired. The data sources are Bureau of the Census (1986, p. 1.4; 1998, p. 1.7) for 1947, 1949–1976; and U.S. Census Bureau (2006, p. 1; 2007) for 1977–2004, 2005–2006. The Census "wages" are really gross earnings, and total hours are an actual-work concept. There are two differences from the Rees AHE series. (1) The Rees 1957 figure is corrected. (2) The series is carried further, to 2006, and in principle is ongoing.

An alternative procedure is to convert the Bureau of Labor Statistics (BLS) AHE from a paid-hour to an actual-work-hour basis via the Hours-at-Work Survey; but the defects of the Survey (see chapter 2, *Hours-at-Work Survey*) mitigate against that procedure. Further, this alternative procedure is indirect and is readily trumped by the adopted Rees and Census foundation of AHE. Also, there are problems with the BLS AHE (see chapter 2, *Current Employment Statistics Survey: Average Hourly Earnings*).

Other alternative series (from table 2.1) are the Employment Cost Index (ECI) and Employer Costs for Employee Compensation (ECEC), and these are inherently on an actual-work-hour basis. The ECI is immediately rejected because it involves fixed weights. The ECEC is a current-weight series and has the advantage of incorporating a corresponding benefits series; but it is purely sample-based and begins some 40 years later than the Census data. Therefore the ECEC advantage is insufficient to drop the Census foundation of the series.

The AHE figure for <u>1948</u> must be estimated. Rees interpolates it using the BLS AHE as interpolator, the logical choice, repeated here; but his method of interpolation is not stated. The method of interpolation followed throughout the present study is as follows. Let Y_i denote the desired series and X_i the interpolator series in year i. Y_0 and Y_n (n > 0) are known; Y_i, i = 1, 2,..., n−1, is to be estimated. Of course, X_i is known for all i, i = 0,..., n. Compute the ratio $R_i = Y_i/X_i$, i = 0, n. R_i, i = 1, 2,..., n−1, is linearly interpolated between R_0 and R_n. Then Y_i, i = 1, 2,..., n−1, is estimated as the product of the linearly interpolated R_i and the known X_i. For the current interpolation, Y is present-study AHE, X is BLS Standard Industrial Classification System (SIC)-based AHE, 0 is 1947 and n 1949. Therefore present-study AHE for 1948 is estimated as the product of "the average of R_{1947} and R_{1949}" (1.0133) and "the BLS SIC-based AHE for 1948." It turns out that the resulting AHE (to two decimal places, that is, to the nearest cent) is identical to the Rees figure.

For <u>1932–1946</u>, the Rees AHE series, again denominated dollars per hour, is taken as is. This segment of the Rees series is composed of two underlying components, 1932–1939 and 1939–1947, that (by construction) seamlessly merge with themselves and with the 1947–1957 segment (see chapter 3, COMPOSITE SERIES), for which the Rees methodology has been adopted. One can wish for richer data availability; but, absent wish fulfillment, it appears impossible to improve on the Rees series for 1932–1946.

The approach to the period <u>1920–1932</u> is different from that of Rees. Rees decidedly rejects the BLS AHE series, in large part because a crucial input—the BLS average-weekly-hours (AWH) series—is biased downward (see chapter 2, *Current Employment Statistics Survey: Average Hourly Earnings* and *Current Employment Statistics Survey*). Instead, Rees employs the National Industrial Conference Board (NICB) AHE series as an interpolator between separately constructed 1890–1920 and 1932–1957 segments, after the 1890–1919 segment is extended to 1920 via the Creamer series (see chapter 3, AVERAGE HOURLY EARNINGS OR HOURLY WAGE RATE). Problems with the Creamer series are plain (varying weighting patterns, wage-rate concept, possible incomplete recording), and therefore that series is not utilized here. Because the level of the NICB series is substantially above both adjacent segments of the Rees series, Rees performs one-year-overlap linking to his 1920 and 1932 AHE and then resorts to linear percentage downward adjustment of the NICB figures for the intervening years. This long interpolation appears excessive. Is there an alternative way of estimating AHE for the 1920–1932 period?

For 1923–1929 (clearly extendible to 1923–1931), Rees acknowledges that the BLS average-weekly-earnings (AWE) series is reliable: "The weekly earnings data, because they use Census [1923 and 1929] bench marks, have reasonably full coverage of small establishments" (Rees, 1960, p. 16). The problem with the BLS AHE (identically equal to AWE/AWH) is a deficient AWH series. For the 1920–1922 AHE, Rees criticizes the BLS adoption of King's (all-worker) data. However, the BLS use of (1919 and 1923) Census relative wage-earner versus salaried-worker earnings to adjust the King figures to a strict wage-earner basis suggests that the BLS-series problem again is not AWE but rather AWH.

Therefore the BLS AWE series is accepted for 1920–1932; but the BLS AWH series must be replaced. Table 5.1 summarizes all the available figures and series for AWH of production workers in manufacturing for 1920–1932. *The hours concept must be actual-work, which eliminates entries of a full-time concept.* Of the remaining entries, some are composed of scattered figures. While one requires a continuous annual series, one would not want to exclude consistent scattered figures, which could serve as benchmark figures. So such entries require examination. Deficiencies of the King series (see chapter 2, *Private Surveys* under DAILY, WEEKLY, OR MONTHLY HOURS OF WORK) make it unsuitable for the present purpose.

The Rees actual-work AWH is computed as the ratio of BLS AWE (Bowden, 1955a, p. 805) to Rees AHE. The numerator series is reliable (see above); but the unsatisfactory state of the denominator series is itself the impetus for the present search! So this Rees, indirect, series is rejected. Kendrick's actual-work series consists of actual and extrapolated BLS figures. However, BLS actual hours are biased downward, which led to the Rees AHE and mandates rejection of the BLS AWH series (final entry in table 5.1).

What remain are the Beney and Jones series. Beney is the NICB series, about which see chapter 2, *Private Surveys* under both EARNINGS AND WAGES and DAILY, WEEKLY, OR MONTHLY HOURS OF WORK. Advantages of the Beney series are the actual-work concept, availability for all years 1920–1932, and computation of annual figures as averages of monthly figures. Disadvantages are underrepresentation of small firms and of the South, fixed-weight pattern, and missing months in 1920 and 1922. The Jones AWH series runs 1900–1957, but only the 1920–1932 segment is of interest here. The Jones series segment is discussed in Jones (1961, pp. 14–22, 128–133; 1963, pp. 377–380), Whaples (1990, pp. 26–28), and Sundstrom (2006e). Whaples (1990, pp. 26–27) judges Jones a "reliable annual

Table 5.1 Average hours per week, 1920–1932: Production workers[a] in manufacturing

Study	Years[b]	Hours Concept	Employment Weights	Method of Construction[c]	Data Source
King (1923, pp. 82, 87)[d]	1920–1922[e]	full-time, actual work	current	direct	surveys
Brissenden (1929, p. 355)	1921	full-time	"	"	COM
Douglas (1930, pp. 112, 114, 116) [all, payroll, and union industries][f]	1920–1926	"	"	"	BLS, CL (1905)
Beney (1936, pp. 44–46)[g]	1920–1932[h]	actual work	constant (year 1923)	"	NICB
Wolman (1938, p. 2)[i]	1920–1932[j]	full-time	"	"	"
Wolman (1938, pp. 2, 8)[k]	1921, 1923, 1929	"	current	"	COM
Bowden (1955a, p. 805)	1923, 1929	"	"	"	"
Rees (1960, pp. 17–18)	1921, 1923–1929	actual work	"	indirect	BLS, NICB
Rees (1960, pp. 17–18)	1921, 1923, 1929	full-time	"	direct	COM
Kendrick (1961, p. 445)	"	full-time, actual work	"	direct[l]	COM, BLS, NICB
Jones (1963, p. 375)[m]	1920–1932	actual work	"	direct	BLS, COM, NICB[n]
BLS Web site[o]	"	actual (1920–1931), paid (1932)	"	indirect (1923–1931), direct (1932), combination (1920–1922)	BLS surveys, COM, King (1923)

Notes:
[a] Also termed "wage-earners."
[b] Annual except where otherwise noted.
[c] Indirect: ratio of AWE to AHE.
[d] Includes nonproduction workers. Reprinted, 1921 annualized, in Rees (1960, p. 18).
[e] Quarterly. 1922 first quarter only.
[f] Reprinted (all, payroll, and union industries) in Bureau of the Census (1960, p. 91; 1975, p. 168) and Sundstrom (2006a p. 2.303). Reprinted (all-industries only) in Bureau of the Census (1949, p. 67).
[g] Reprinted in NICB (1950, pp. 340, 342), among other NICB publications; in Bureau of the Census (1960, p. 94; 1975, p. 172), and in Fabricant (1942, p. 234). Reprinted, 1921 and 1923–1929, in Rees (1960, pp. 17–18).
[h] Average June–December for 1920; average July–December for 1922; other years, 12-month average.
[i] Also in various NICB publications; for example, NICB (1930, p. 51), quarterly.
[j] Average of January–March for 1932.
[k] Reprinted in Whaples (1990, p. 34).
[l] But see "BLS Web site" entry.
[m] Reprinted in Shiells (1985, p. 22), Whaples (1990, p. 34), and Sundstrom (2006e).
[n] Also, Magdoff, Siegel, and Davis (1939), Fabricant (1942), and annual reports of two iron-and-steel firms. See Jones (1961, pp. 129–130).
[o] www.bls.gov.

CL = Commissioner of Labor.
COM = Census of Manufactures.
NICB = National Industrial Conference Board.

series...superior to the frequently-used Bureau of Labor Statistics (BLS) series which gives hours paid for, not hours worked, and surveys only larger firms."

The Jones (actual-work) AWH series is based on especially constructed benchmark figures for 1929 and 1933. The 1929 benchmark figure is derived as a two-step process. First, the Census-employment-weighted mean of AWH for 27 industries is computed, industry hours mainly from NICB and BLS surveys. Second, the weighted mean is multiplied by the Census-scheduled-hours ratio of all manufacturing to the industries included in the weighted mean. The 1933 benchmark figure is the product of the BLS all-manufacturing AWH and the Rees adjustment ratio (employment-weighted average of ratio of Census Man-Hour-Statistics AWH to BLS AWH, for all manufacturing industries common to both samples—see Rees, 1960, p. 11). Rees applies this procedure to the revised BLS all-manufacturing AWH figure. Jones (1963, p. 378, note 9) enhances consistency by using the unrevised figure, in the same BLS bulletin from which Rees obtains the individual-industry AWH to construct the adjustment ratio.

For 1920–1928, the NICB series is used as an extrapolator from the 1929 benchmark figure. The NICB figures for 1920 and 1922, based on averages of seven and six months, are adjusted to full-year averages using data from King—Jones is as thorough as can be. For 1930–1932, the NICB series and BLS all-manufacturing series are used to interpolate between the 1929 and 1933 benchmark figures.

The only serious criticism of the Jones series is the fixed-employment-weights property of the NICB series. Jones (1963, p. 378, note 11) notes that, as a consequence, the NICB series "tends to overstate the effect of the decline in hours in durables upon the all-manufacturing average of hours of work." Fortunately, at least according to Jones, this is an issue only for 1932–1933, "due to the earlier decline, during the depression, of both hours of work and employment, in the durable goods sector of manufacturing relative to the nondurable."

Therefore it is with confidence that the 1920–1931 segment of the Rees AHE is replaced by a series constructed as follows. First, for 1920–1932, an indirectly computed AHE is the ratio of SIC-basis AWE (dollars per week; source: BLS website) to AWH (average hours of work per week; source: Jones, 1963, p. 375), for production workers in manufacturing. Second, the 1932 ratio of the new AHE (also the Rees AHE) figure to this indirectly computed AHE is calculated as 0.9922, happily close to unity. Third, for 1920–1931, the new AHE is the product of this ratio and the indirectly computed AHE. Thus the entire 1920–2006 segment of the new AHE series is developed.

1800–1919

Benchmark Figures

Average Annual Earnings

Table 5.2 exhibits existing adjusted Census average annual earnings (AAE) for 1859–1919. For the 1889–1919 period, there exist two internally consistent segments of corrected Census-based AAE of wage-earners in manufacturing. These are Douglas and Rees, shown as the first two entries in table 5.2. For each Census year, AAE is computed as the ratio of total wage payments (earnings) to the average number of employed wage-earners. Adjustments, summarized in the table, are made either to the ratio directly (Douglas) or to numerator and denominator separately (Rees), as follows.

First, construction and other hand and/or custom trades are included in manufacturing in 1889; but Census data for 1899 exist both including and excluding these trades, thus enabling computation of ratios to exclude also the trades in 1889. Douglas uses only one, the all-industries, ratio for this purpose, whereas Rees makes the correction industry-by industry, resulting in enhanced consistency of the revised 1889 AAE with the later-Census AAE.

Second, while Douglas accepts the definition of manufacturing industries current for each Census, Rees adopts a uniform, consistent, later definition, that of the 1939 Census. Therefore, while Douglas does not omit any (non-construction, non-hand-trade) industry, Rees excludes all industries not in the 1939 Census (with the exception of one industry, "tinplate and terneplate"). A list of "obsolete" manufacturing industries in the 1899–1937 Censuses, with the associated number of wage-earners (but not wages) is in Fabricant (1942, pp. 213–214).

Two further adjustments are made by Rees but not Douglas. Third, logging establishments, included in the lumber industry only from 1899 onward, are incorporated in 1889 via appropriate included/excluded ratios for 1899. Fourth, for the 1909, 1914, and 1919 Censuses, the wage-earner (not the wages) figure for seasonal industries is corrected for downward bias, given that the Rees (and present) methodology requires average employment on a time-of-plant operation rather than calendar-year basis. The glass industry is mentioned as an example of a seasonal industry, but the method of correction is not described.

Obviously, the Rees AAE figures have greater uniformity and consistency than the Douglas figures. Therefore, notwithstanding the incomplete description of seasonal adjustment, it is an easy decision to adopt the Rees data for the Census years 1889, 1899, 1904, 1909, 1914, and 1919. Fortunately, Kendrick (1961, pp. 438–440) provides a full description of a related seasonal adjustment. Kendrick makes adjustments similar to those of Rees, but with three differences: (1) Adjustments are for 1869–1889. (2) The overall concern is only with wage-earners, not wages. (3) The seasonal adjustment is of an opposite nature: from a time-of-plant-operation to a calendar-year basis. It is likely that the Rees seasonal adjustment is closely related to that of Kendrick, because Rees (1961, p. 30, note 13) does refer to the table listing all the adjustments in the (then forthcoming) Kendrick study.

For the 1849–1879 Censuses (extended to 1889, for overlap and linking), there are several alternatives for constructing AAE. First, one could apply the Rees 1889–1919 methodology. This procedure is inadvisable, because the ratios for 1899 inherent in Rees' adjustments for partly hand and/or custom trades and for logging become increasingly obsolete as one goes back in time earlier than 1889. Also, for some industries, the 1849–1879 period has the problem of intermittent reporting, concerning which Rees provides no guidance.

Second, one could adopt the Long figures for 1859–1889, shown as the last entry in table 5.2 and described in Long (1960, pp. 40–42); but there are problems.

(1) While Long excludes industries that are entirely hand and/or custom trades (1859–1889) as well as industries classified as mining, forestry, fishing, and agricultural processing (1859 only), he does not enumerate the individual industries involved. True, Long provides references to lists of Census hand trades in North (1899, p. 271) and Easterlin (1957, p. 642). However, neither North nor Easterlin makes adjustments for 1859. The present author could not replicate Long's result for that year, wherefore it is impossible to extend Long's procedure consistently to 1849.

(2) Long omits two industries—boots and shoes, men's clothing—on grounds that they embody "large numbers of custom and repair shops." Yet women's clothing is retained.

(3) Long's treatment of "intermittently reported manufactures" is questionable. His procedure apparently is to exclude, for all Census years, any industry that is not reported in at least one Census year over 1859–1889; but his adjustment figures omit years during which some of his identified industries are in the Census (for example, "gas,

illuminating," "grindstones and millstones," "flax-dressing," and "cars, omnibuses and repairing" in 1859). In effect, though inconsistently, Long is omitting some industries later considered nonmanufacturing.

(4) The exclusion of "gas, illuminating and heating" is particularly inconsistent, as "illuminating gas" is included in Long's Aldrich-based wage series.

The third approach, adopted here, errs on the side of inclusion rather than exclusion of industries and provides a consistent treatment for the entire 1849–1889 period. The definition of manufacturing in the 1904 Census (Bureau of the Census, 1907, pp. xxii–xxiv, 3–20) is adopted. Rees' 1939 Census definition is too far removed from 1849–1879. Nonmanufacturing industries and pure hand and/or custom trades are excluded, with Easterlin's list, along with the 1904 Census, utilized for this purpose. In the included industries, neighborhood (non-factory) establishments are retained along with their factory counterparts—another manifestation of inclusiveness. As one goes back in time, neighborhood establishments have greater importance relative to factory establishments. It is inappropriate to adopt a narrow (exclusively factory) definition of manufacturing for the early period. (Also, *both* series to extend the 1820, 1831, 1849 AAE from a Northeast adult-male to a U.S. all-worker basis *and* series to interpolate AAE between benchmark years include hand and custom trades—see **Interpolator and Extrapolator Series**.) Rather than excluding intermittently reported industries, the opposite is done: interpolation for missing years is used to achieve continuity. Industries with large custom-and-repair components are included in their entirety, with data lacking to omit the repair component (the custom component would be retained in any event).

Finally, there is complete transparency in the procedure. Table 5.3 exhibits the hand and custom trades eliminated from the 1849–1889 figures, and table 5.4 shows the trades retained. Tables 5.5 and 5.6 provide details of the computation procedure to obtain adjusted AAE for 1849, 1859 and 1869, 1879, 1889, respectively.

The 1849–1889 AAE so constructed is made consistent with the 1889–1919 present-study AAE via using the 1889 overlap (ratio of 1889 AAE on 1889–1919 basis to 1889 AAE on 1849–1889 basis) as a multiplicative factor. The 1889 ratio is 0.9720, pleasingly close to unity. Thus the all-worker, all-U.S. AAE has been assembled for Census, benchmark, years 1849 to 1919.

For the pre-1849 period, there are two sources of benchmark wage-rate data: Lebergott versus Sokoloff and Villaflor [SV], summarized in table 5.7. Lebergott (1964, pp. 285–289) is the only discussant

Table 5.3 Hand and custom trades eliminated from 1849–1889 Census figures

Trade	Last Census Reported
bakers	1849
bicycle and tricycle repairing	1889[a]
blacksmithing	1899
bleaching and dyeing	1869
bottling	1899[b]
butchering	1869[c]
carving	1859
coffins and burial cases, trimming and finishing	1889[a]
cotton cleaning and rehandling	1889[a]
cotton compressing	1899[d]
cotton ginning	1899[e]
dyeing and cleaning	1899[f]
dyers	1849
kindling wood	1899[g]
lock and gun smithing	1899[g,h]
photography	1899[g]
rigging	1859
taxidermy	1899[i]
tobacco, stemming and rehandling	1899[f]
upholstery	1869
watch, clock, and jewelry repairing	1899[g,j]
wheelwrighting	1899[k]
whitesmithing	1859[l]
wood cutting and cording	1859

Notes:
[a] Not reported in earlier years.
[b] Not reported in 1849, 1859, 1879.
[c] Not reported in any other year.
[d] Not reported in 1849, 1869.
[e] Not reported in 1849, 1869, 1879.
[f] Not reported in 1849–1869.
[g] Not reported in 1849.
[h] Reported as "locksmithing and bellhanging" in 1859–1869. "Gunsmithing" reported as separate industry in 1869. "Locksmiths" included in "white and lock smiths" in 1849.
[i] Not reported in 1849–1859.
[j] Reported as "watch and clock repairing" in 1869; estimated component of "watches, watch repairing, and materials" in 1859.
[k] Not reported in 1859.
[l] Reported as "white and lock smiths" in 1849.

Table 5.4 Hand and custom trades retained in 1849–1889 Census figures

Trade	Censuses Reported	Reason for Retention
boots and shoes: custom work and repairing	1879–1899	presumed included in "boots and shoes" 1849–1869
cheese and butter, urban dairy product	1889–1899[a]	presumed included in "cheese, butter, and condensed milk" 1879, "cheese" and "milk, condensed" 1859, "cheese" 1849 and 1869
clothing, men's: custom work and repairing	1889–1899[b]	presumed included in "clothing, men's" 1859–1879, "clothiers and tailors" 1849
clothing, women's: dressmaking[c]	1889–1899	presumed included in "clothing, women's" 1869, "clothing, ladies'" 1859, "clothiers and tailors" 1849
furniture, including cabinetmaking: repairing, and upholstering	1879–1899[d]	presumed included in furniture (various classifications) 1869, "furniture, cabinet, school, and other" 1859, "cabinet ware" 1849
millinery, custom work	1889–1899	presumed included in "millinery and lace goods" 1879, "millinery" 1869, "millinery and dressmaking" 1859, "milliners" 1849
sewing-machine repairing	1889–1899	presumed included in various sewing-machine classifications 1859–1879,[e] not reported 1849

Notes:
[a] Separable component of "cheese, butter, and condensed milk," via individual-state data.
[b] Separable component of "clothing, men's," via individual-state data.
[c] Only trade listed that is unquestionably excluded from a Census—in this case, the 1879 Census. Impossible to interpolate in a reasonable way.
[d] Separable from "furniture, factory product" in 1889 and 1899.
[e] "Included in other classifications in 1880 [1879]" (U.S. Census Office, 1902a, p. 14, note 4).

Table 5.5 Computation of average annual earnings, manufacturing: 1849 and 1859

Item	1849		1859	
	Wage-earners (number)	Wages (dollars)	Wage-earners (number)	Wages (dollars)
All industries: Census	966,969	236,745,858	1,311,246	378,878,966
Additions				
gilding[a]	—	—	109	52,242
upholstery materials[b]	183	50,851	172	59,179
weaving[a]	—	—	280	51,330
Deductions				
nonmanufacturing sectors				

Continued

Table 5.5 Continued

Item	1849		1859	
	Wage-earners (number)	Wages (dollars)	Wage-earners (number)	Wages (dollars)
agriculture[c]	—	—	146	42,088
fisheries	21,238	4,639,188	30,383	6,077,677
forestry[d]	957	179,972	1232	327,472
mining[e]	23,024	8,498,814	90,340	36,361,430
quarrying[f]	10,081	3,452,722	370	118,824
construction[g]	16,729	6,095,188	12,999	5,375,448
services[h]	—	—	405	140,496
hand and custom trades				
bakers	6727	1,960,416	—	—
blacksmiths	25,002	6,508,032	15,720	4,827,303
carving	168	78,516	229	105,596
cotton ginning	—	—	271	52,644
cotton pressing	—	—	64	25,920
dyeing and bleaching	565	161,688	3203	953,024
dyers	460	127,320	—	—
kindling wood	—	—	416	131,892
locksmithing and bellhanging	—	—	203	76,992
photographs	—	—	653	359,854
rigging	57	31,464	294	147,588
upholstery	1512	365,580	1427	425,452
watch and clock repairing[i]	—	—	434	92,588
wheel wrights	11,549	3,157,544	—	—
whitesmithing[j]	415	140,712	9	3504
All industries: adjusted	848,668	201,399,553	1,153,009	323,395,925
Average annual earnings[k] (dollars)		237		280

Notes:

[a] Linearly interpolated via 1849 and 1869 figures. Provides continuous coverage of industry, 1849–1869.

[b] Computed as product of 1869 (upholstery-materials/upholstery) ratio and desired-year upholstery figure. Provides continuous coverage of industry, 1849–1889.

[c] Grain-thrashing, husks (prepared), clover-hulling and seed-cleaning, hay-pressing, prepared moss, seeds (garden and flower).

[d] Timber-cutting and timber hewed (timber-hewers), wood-cutting (and cording).

[e] Barites, chrome, clay, coal, coke, copper, corundum, emery, gold, iron ore, isinglass (mica), lead (mining and smelting), magnesia, manganese, nickel ore, ocher, plumbago (black and silver lead), silver, sulfur, zinc ore.

[f] Slate, stone, and marble.

[g] Carpenters and builders (carpentering), plumbing and gas-fitting (plumbers), stair-building, stucco and stucco work, painting, plastering, roofing, bridges, cement (for building purposes).

[h] Dentistry and laundry-work.

[i] Computed as product of 1869 ratio ("watch and clock repairing")/("watch and clock repairing" plus "watch materials" plus "watches") and 1859 "watches, watch repairing, and materials."

[j] "White and lock smiths" in 1849.

[k] Ratio of wages to wage-earners.

Source: All data from Walker (1872, pp. 399–408).

Table 5.6 Computation of average annual earnings, manufacturing: 1869, 1879, and 1889

Item	1869		1879		1889	
	Wage-earners (number)	Wages (dollars)	Wage-earners (number)	Wages (dollars)	Wage-earners (number)	Wages (dollars)
All-industries: Census	2,053,996	775,584,343	2,732,595	947,953,795	4,251,613	1,891,228,321
Additions						
rice cleaning and polishing[a]	479	95,152	—	—	—	—
stationery[b]	1755	633,137	—	—	—	—
cars and general shop construction and repairs by steam railroad companies[c]	—	—	64,624	38,718,612	—	—
gas: illuminating and heating[d]	—	—	10,860	7,523,180	—	—
Deductions						
nonmanufacturing sectors						
agriculture[e]	—	—	—	—	787	144,870
construction[f]	101,262	42,771,898	105,685	47,691,809	355,402	220,214,702
services[g]	1020	184,272	541	269,044	1486	768,401
utilities[h]	—	—	229	117,500	1794	1,285,242
hand and custom trades						
bicycle and tricycle repairing	—	—	—	—	212	102,141

Continued

Table 5.6 Continued

Item	1869		1879		1889	
	Wage-earners (number)	Wages (dollars)	Wage-earners (number)	Wages (dollars)	Wage-earners (number)	Wages (dollars)
blacksmithing	52,982	9,246,549	50,634	16,200,800	26,539	13,499,738
wheelwrighting	6989	1,353,474	—	—	3044	1,511,678
bottling	89	28,470	—	—	—	—
butchering	1881	546,346	—	—	1826	946,441
coffins and burial cases, trimming and finishing	—	—	—	—	—	—
cotton, cleaning and rehandling	—	—	—	—	196	43,248
cotton compressing	—	—	1008	573,005	2785	1,040,808
cotton ginning	—	—	—	—	6920	650,885
dyeing and cleaning[i]	4172	1,783,449	1467	511,886	3991	1,717,387
gunsmithing	1082	228,879	—	—	—	—
kindling wood	701	253,150	1443	526,861	1617	616,094
lock and gunsmithing[j]	555	160,799	887	368,967	1242	655,861
photography[k]	2800	786,702	3977	1,751,118	6967	3,472,643
taxidermy	18	5700	150	22,000	90	46,362
tobacco, stemming and rehandling	—	—	1534	170,871	5985	1,128,517
upholstery	4757	1,679,217	—	—	—	—
watch, clock, and jewelry repairing[l]	2025	459,492	1657	866,996	4819	2,861,120

Continued

Table 5.6 Continued

Item	1869		1879		1889	
	Wage-earners (number)	Wages (dollars)	Wage-earners (number)	Wages (dollars)	Wage-earners (number)	Wages (dollars)
All-industries: adjusted	1,875,897	716,824,235	2,638,867	925,124,730	3,825,911	1,640,522,183
Average annual earnings[m] (dollars)		382		351		429

Notes:

[a] Linearly interpolated via 1859 and 1879 figures. Provides continuous coverage of industry, 1859–1889.

[b] Linearly interpolated via 1859 and 1879 figures. Provides continuous coverage of industry, 1849–1889.

[c] First, sum of "'cars and general shop construction and repairs by steam railroad companies' and 'cars, railroad and street, and repairs, not including establishments operated by steam railroad companies'" linearly interpolated via 1869 and 1889 figures. 1869: "cars, railroad and repairs" comprises "car-repairing," "cars, railroad, horse," and "cars, railroad, steam"; components not reported separately in Census, notwithstanding contrary statement in Walker (1872, p. 394, note d). 1889: sum of "cars and general shop construction and repairs by steam railroad companies" and "cars, railroad and street, and repairs, not including establishments operated by steam railroad companies." Second, 1879: "Cars and general shop construction and repairs by steam railroad companies" computed as interpolated sum "'cars and general shop construction and repairs by steam railroad companies' and 'cars, railroad and street, and repairs, not including establishments operated by steam railroad companies'" minus reported "cars, railroad ard street, and repairs, not including establishments operated by steam railroad companies." Provides continuous coverage of "cars and general shop construction and repairs by steam railroad companies" and "cars, railroad and street, and repairs, not including establishments operated by steam railroad companies" industries with corresponding industries, 1849–1889.

[d] Linearly interpolated via 1869 and 1889 figures. Provides continuous coverage of industry, 1849–1889.

[e] "Hay and straw baling" and "teasels."

[f] Carpentering (and building), plumbing and gas-fitting, bridges (bridge-building), plastering, masonry (brick and stone), painting, paper-hanging, paving and paving materials, roofing and roofing materials, cement.

[g] Dentistry.

[h] Electric light and power.

[i] "Bleaching and dyeing" in 1869.

[j] "Locksmithing and bellhanging" in 1869.

[k] "Photographs" in 1869.

[l] "Watch and clock repairing" in 1869.

[m] Ratio of wages to wage-earners.

Source: All data from Walker (1872, pp. 394–398) and U.S. Census Office (1902a, pp. 3–17).

Table 5.7 Average wage rate in antebellum period—Northeast[a], wage-earners in manufacturing

Study	Years	Denomination	Workers	Data Source
Lebergott (1964, pp. 289, 547)	1831[b]	dollars per day	all	primarily McLane Report
Sokoloff and Villaflor (1992, p. 36)[c]	1820, 1831, 1849, 1859	dollars per year[d]	adult males	Census of Manufactures and McLane Report schedules

Notes:
[a] New England (Connecticut, Maine, Massachusetts, New Hampshire, Rhode Island, Vermont) and Middle Atlantic (New Jersey, New York, Pennsylvania).
[b] Misstated as 1832.
[c] Reprinted, for New England and Middle Atlantic separately, in Margo (2006c).
[d] For 1831, annualization of (monthly, weekly, daily) wages based on assumption of (12 months, 52 weeks, 310 days) of employment per year. For 1849 and 1859, annualization of monthly wages. See Sokoloff and Villaflor (1992, pp. 35–36).

of his work, while the SV (1992, pp. 31, 35–39) own presentation is supplemented in Sokoloff (1982, pp. 287–294; 1986, p. 686), Goldin and Sokoloff (1982, p. 753), and Margo (2006c). Which source should be adopted? Both have the advantage of pertaining to wage-earners in manufacturing—exactly what is required. Both, pleasingly, weight firm and industry wages by employment; but both, unpleasingly, are confined to the Northeast (the nine states of the New England and Middle Atlantic regions—see table 5.7, note a). Both studies utilize the McLane Report, with all its limitations (see chapter 2, *Congress and Treasury* under EARNINGS AND WAGES).

Lebergott uses primarily the McLane Report (1833), with some other information, to compute an all-manufacturing average of 12 individual-industry wage rates. Using "all available reports for each state as presented in the 1,900 pages of the McLane Report" (with the exception of the, heavily overrepresented, shoe industry in Massachusetts, for which the smallest firms are dropped), Lebergott (1964, p. 285) mines the Report with a thoroughness equaled by no other scholar before or after. SV have more industries (18); but they base their figures on samples of the schedules of firms in the Censuses and Report rather than on a complete count.

SV restrict their demographic group to adult males, whereas Lebergott incorporates males and females of all ages. Also, Lebergott alone deserves credit for making an attempt (although minor) to extend source data beyond the Northeast. SV compute average annual wages, while Lebergott calculates an average daily wage rate. Here Sokoloff and Villaflor are superior, because the objective is to link the antebellum benchmark figures to the AAE figures of the postbellum era.

Both studies must make arbitrary assumptions. Lebergott uses data of varying quality and exercises subjective judgment on wage estimates and weighting patterns. SV make assumptions to annualize wage data of higher than annual frequency and to restrict demographic coverage to adult males. For the former, see table 5.7, note d. Regarding the latter: (1) For 1820, the Census annual wage bill is not separated by age or sex. SV assume that females and children earned 35 percent of the adult-male wage. (2) For 1849 and 1859, Census data are for all males. SV assume that, for each industry, boys constituted the same proportion as in 1820. Further, for industries in which the boy proportion exceeded 33 percent in 1820, SV assume that the proportion fell to 33 percent in 1849 and 1859. Also, the boy wage is assumed to be half the adult-male wage.

The deciding factor to select the SV (1820, 1831, 1849) figures—denoted as $AAE(SV)_{AM,NE}$ (average annual earnings of adult males in the Northeast, Sokoloff-Villaflor data)—as benchmarks is that they provide both an additional antebellum benchmark year (1820) and an overlap (1849) with the previously selected (Census-years 1849–1919) AAE figures. Of course, it will be necessary to transform the SV figures from adult-males to all workers and from the Northeast to the entire country; this is done at the average-daily-earnings (ADE) level—see *Average Daily Earnings*.

Days of Operation
In order to generate average full-time daily earnings (on the way to constructing AHE), AAE must be divided not by the number of calendar days in the year but rather by the average number of days of operation (ADO) of manufacturing establishments. Table 5.8 presents the figures for ADO that have been generated by previous authors, and the figures used in the present study (last column). Using the 1904 Census frequency distribution of number of establishments according to number of days of operation (see chapter 2, DAYS OF OPERATION), Rees (1960, pp. 34–35) computes a U.S. average (and also a three-state average), on the assumption that the mean of each interval is the midpoint (except that he selects 20 days as the mean of the 0–30 day interval). The resulting figures are shown in the second and third columns of the table.

Table 5.8 Average number of days of operation—Manufacturing, Census years, 1849–1919

Year	Rees					Atack and Bateman[d] (United States)	Atack, Bateman, and Margo (United States)	Present Study[e] (United States)
	United States[a]	Three States[b]		Long				
		Census Data	State Data	United States	Four States[c]			
1849	—	—	—	—	—	—	—	276
1859	—	—	—	286[f,g]	—	—	—	287
1869	—	—	—	246[f,g]	—	262[f]	277[f,h]	277
1879	—	—	—	257[f,g]	—	272[f]	272[f,h]	272
1889	—	—	—	279[f,i]	260[j]	—	—	267
1899	—	—	290	—	—	—	—	275
1904	263	277	288	—	—	—	—	273
1909	—	—	289	—	—	—	—	274
1914	—	—	281	—	—	—	—	267
1919	—	—	279	—	—	—	—	265

Notes:
[a] Census data.
[b] Massachusetts, New Jersey, Pennsylvania.
[c] Iowa, Maine, Maryland, Ohio.
[d] Superseded by Atack, Bateman, and Margo.
[e] See text.
[f] Year in source construed as Census nominal year (one-year later than shown in first column).
[g] Computed indirectly as ratio of average annual earnings to average daily wage.
[h] Employment-based full-time-equivalent months of operation multiplied by 25.75, see text.
[i] Based on Census unemployment data, see text.
[j] Pertains to 1890. "Median of 260 days computed from weighted averages for four states" (Long 1960, p. 48, note 14).

Source: Atack and Bateman (1995, pp. 2, 12); Atack, Bateman, and Margo (2002, p. 797), Long (1960, p. 48), Rees (1960, p. 19; 1961, pp. 33, 35). Also, see text.

Rees offers two reasons why the Census-based averages are biased downward. First, because the distribution is "markedly skewed to the left," the errors involved in taking midpoints are not entirely offsetting and, therefore, the overall mean is underestimated. Second, the frequency distribution counts each establishment as one unit, irrespective of size (number of workers). In general, large establishments have more days of operation than smaller firms. (Rees reports on this relationship; for systematic positive evidence, see Atack, Bateman, and Margo ([ABM], 2002, pp. 801, 803). For consistency with the AHE concept of the present study, establishment (or industry) components of ADO should be weighted by employment. Therefore, from this standpoint, the computed average overweights small firms and understates the number of days of operation.

Using state-labor-bureau data, Rees computes a three-state (Massachusetts, New Jersey, Pennsylvania) ADO series (see chapter 2, DAYS OF OPERATION), shown for Census years in the fourth column of table 5.8. Rees uses employment weights to combine figures both for industries within states and for the three states. That is pleasing; but, as Douglas (1962, p. 447) implies, to ignore the 1904 Census figure (as Rees does) discards data. The solution adopted here is to convert the Rees state-data-based three-state figures to a national basis, via the multiplicative factor 263/277 (the Census-based U.S./three-state ratio for 1904). Resulting figures, for the Census years 1899–1919, are in the final column of table 5.8. Applying this method to the year 1889 would be unwise, because prior to 1892 Massachusetts alone has usable data.

Using "nationally representative [samples]...of the universe of surviving manuscript returns from the two [1870 and 1880] censuses," ABM (2002, p. 794) present figures for full-time equivalent months of operation with establishments weighted alternatively by the value of capital stock and by employment. Months may be converted to days using the multiplicative factor of 25.75 days per month (Atack and Bateman, 1992, p. 154; 1995, p. 5; ABM, 2002, p. 798, note 19; 2003, p. 179). The employment-weighted figures are the more-appropriate for the present study, and the pertinent figures are in the next-to-last and last columns. The authors neglect to mention that the Census years 1870 and 1880 translate effectively into the calendar years 1869 and 1879 (see chapter 2, *Regular Census* under EARNINGS AND WAGES). The, superseded, Atack-Bateman (1995) figures are shown in the seventh column. Only the 1869 figure is changed, but that substantially. The samples utilized by ABM—composed of 5296 and 8173 firms in 1870 and 1880—are discussed in Atack and Bateman (1999, pp. 180–185) and ABM (2002, p. 798). Some of the industries in the sample (agricultural services, carpentry, blacksmithing) are not clearly in the manufacturing sector.

Long (1960) uses Census unemployment data and the Census figure for number of workers attached to manufacturing to generate an unemployment rate of 6.6 percent for 1889 (1890 in his reckoning). Applying that rate to a full-time year (days of operation, which he calls days of employment) of 299 days, the result is an average of 20 idle days and therefore a 279 ADO. Long corrects his Aldrich-data average daily wages (ADW) for level so that the product of adjusted wages and ADO yields his predetermined AAE figure. Applying the corrective factor to ADW for the other Census years of his study, ADO is obtained as the ratio of AAE to adjusted ADW. Long's ADO figures are shown in the fifth column of table 5.8. He also uses days-worked, months-lost, and days-lost data for four states (different from the Rees states) to obtain an alternative ADO figure for 1890 (sixth column), but about which he is skeptical.

A minor criticism of Long's ADO figures is the lack of synchronization between ADW and the true applicable Census year. Also, both his AAE and ADW figures have deficiencies (see *Average Annual Earnings*; and chapter 3, AVERAGE DAILY WAGE RATE). The most-serious weakness is the application of ADW to estimate ADO, whereas the technique of the present study involves the opposite relationship! While it would be illogical for the Long ADO figures to be inputs into the present-study ADO, they can serve as a check.

ADO for 1849, 1859, and 1889 remains to be obtained. The technique used is based on the assumption that ADO (in manufacturing) is positively and closely related to the position of the manufacturing sector in the business cycle. Let V denote value-added in manufacturing, the product of a smoothed or trend component (VS) and a cyclical component (VC): $V = VS \cdot VC$, whence $VC = V/VS$. Then ADO for missing years is estimated as the product of VC for these years and the average ADO/VC ratio for selected known years.

Value-added in manufacturing (V) for 1839, 1849, 1859, 1869, 1879, 1889, 1899 is taken from Gallman (1960, p. 43). Gallman uses Census data (plus some other sources for 1839), eliminating nonmanufacturing industries and hand trades (oddly, retaining wheelwrighting), estimating figures for omitted or poorly returned industries, correcting an overreported (textile) industry in 1879, and extrapolating from 1879 via Easterlin (1957, p. 694) figures for 1889 and 1899. Gallman's figures for five other years within 1839–1899 are the result of interpolation, and therefore not used here.

To estimate value-added in intercensal years, the Davis (2004, p. 1189) annual industrial-production index is used for 1839–1899. This series is far superior to alternative indexes—the Persons (1931, pp. 170–171) and Frickey (1947, p. 54) indexes of the production of manufactures, the Miron and Romer (1990, p. 336) index of industrial production—in time period, nature of component series, weighting pattern, and behavior.

1. The Frickey series begins in 1860, Persons in 1863, and Miron-Romer in 1884. Thus these series are useless in interpolating Gallman so that ADO for 1849 and 1859 is estimatable.

2. Final products, versus intermediate goods and raw materials, constitute a much higher proportion of the Davis index than they do of the alternatives. (See Davis, 2004, p. 1194, for a direct comparison with Frickey and Miron-Romer; see also Persons, 1931, pp. 175–176.)

3. The Persons and Frickey indexes are less reliable for the early years than for the later period (see, for example, Persons, 1931, p. 173; Frickey, 1942, p. 180; 1947, p. 3; Calmoris and Hanes, 1994, p. 412). Davis (2004, pp. 1186–1187) mitigates this problem by applying both an antebellum and a postbellum weight base.

4. The Frickey series is found to be "excessively volatile" (Romer, 1986, p. 325), indicative of exaggerated cyclical fluctuations in size and frequency. This finding is corroborated in Davis (2004, pp. 1193–1194), in a comparison of Frickey with his own series.

So a continuous value-added series, 1839–1899, is obtained by using the Davis series to interpolate between the Gallman Census-year figures. The interpolation method is that established in 1920–2006, with Y the Gallman figures, X the Davis series, and (1839, 1849), (1849, 1859),…, (1889, 1899) the (0, n) pairs. Thus the Gallman V is extended to a continuous annual series.

The next step is to decompose V into trend (VS) and cyclical (VC) components. The Hodrick-Prescott (HP) filter is used for this purpose. The case for Hodrick-Prescott is summarized by Ravn and Uhlig (2002, p. 371):

> Although the use of the HP filter has been subject to heavy criticism…it has withstood the test of time and the fire of discussion remarkably well. Thus, although elegant new bandpass filters are being developed,…it is likely that the HP filter will remain one of the standard methods for detrending.

The HP filter selects the VS_t series, $t = 1,…, T$, as the solution to minimizing

$$\Sigma (V_t - VS_t)^2 + \lambda \cdot \Sigma (\Delta VS_{t+1} - \Delta VS_t)^2$$

where the first summation is over $t = 1,…, T$; the second summation is over $t = 2,…, T-1$; $\Delta VS_j = VS_j - VS_{j-1}$; and λ is a positive "smoothing parameter," the value of which is predetermined. In the HP equation, $V_t - VS_t$ is sometimes called the "business-cycle component" and $\Delta VS_{t+1} - \Delta VS_t$ may be termed the "trend-accelerator component." The relative importance of the trend-accelerator component decreases (increases) as λ becomes smaller (larger). As $\lambda \to 0$, the "solved VS_t" $\to V_t$ (for all t) and the "resultant VC_t" $\to 1$, meaning there is no cycle, resulting from a trend of minimum smoothness.

As $\lambda \to \infty$, $VS_t \to a + b \cdot t$ (a linear trend), and $VC_t \to V_t/(a + b \cdot t)$—a cycle given maximum scope resulting from a trend of maximum smoothness.

Note that, for purpose of estimating ADO, the cycle (VC_t) is defined not as $V_t - VS_t$ but rather as V_t/VS_t (see above). So "no cycle" ($V_t = VS_t$, all t) involves a VC of unity rather than zero. The reason for the unconventional definition of VC is that it gives rise to a logical estimator of ADO.

Obviously, the resulting VC—and therefore the estimator of ADO—depends heavily on the assigned value of λ. For annual data (as in the present study), values of 10, 100, and 400 have been suggested in the econometrics literature. However, Ravn and Uhlig (2002) make a strong case for a value of only 6.25, and their recommendations is followed here. One is aware that λ so low results in a cyclical component, VC, contained within a narrow band around unity. The consequence is a low variability of the resulting estimated ADO. That makes economic sense; one would expect manufacturing industry as a whole to experience ADO that does not wildly fluctuate.

The ratio $RAY_t = ADO_t/VC_t$ is known for three years: t = 1869, 1879, 1899. It is reasonable to use all available information to estimate the ADO for the earliest Census years, 1849 and 1859. Thus estimated $ADO_t = \{(RAY_{1869} + RAY_{1879} + RAY_{1899})/3\} \cdot VC_t$, t = 1849, 1859. Because the missing ADO_{1889} is flanked by known RAY_t in adjacent Census years, estimated $ADO_{1889} = \{(RAY_{1879} + RAY_{1899})/2\} \cdot VC_{1889}$.

The resulting ADO figures are shown in the pertinent rows of the final column of table 5.8. The relatively high ADO (287) for 1859 appears out of line with ADO for other years. Consider, though, that the figure of 287 is very close to Long's estimate of 286, which is obtained independently for the same year. On the other hand, Long's figure of 246 for 1869 is much too low, compared to figures for any other year and of any other author. Long's method fails for the year 1869. Also, it may be that his figures for 1879 and 1889 differ from those of the present study because of a timing factor. Consider that the ratio $ADO_{1859}/\{(ADO_{1879} + ADO_{1889})/2\}$ is 1.067 for Long and

1.065 for the present study—very close indeed. One concludes that the three self-estimated ADO figures in the final column of the table are sufficiently reliable to proceed.

Average Daily Earnings
Having obtained both AAE and ADO, nationwide, for all Census (and benchmark) years 1849–1919, average full-time daily earnings (ADE) are the ratio AAE/ADO. However, is the description of the computed ADE correct? Assuming that AAE and ADO are measured without error, then ADE is precisely average daily earnings. The issue is whether ADE also warrants the designation *full-time*. ABM (2002, pp. 804–807), using evidence from the 1869 and 1879 Censuses, show that "workers in part-time establishments, on average, received higher monthly wages than workers at full-year establishments" (ABM, 2002, p. 793). They demonstrate that the reason is not differences in the skilled-versus-unskilled composition of the workforce—which suggests, rather, a compensating premium for unemployment risk. The implication here is that full-time ADE may be overestimated.

The ABM finding would not be serious for this study, if the part-time/full-time ratio did not vary significantly over time. However, ABM (2002, pp. 792–793) note "the conventional wisdom that part-year operation in American manufacturing declined in significance over the course of the nineteenth century." Their own evidence shows that "for the median (or typical) manufacturing establishment, full-year operation was already the norm as early as 1870" (ABM, 2002, p. 796). It is likely that the part-time/full-time ratio was relatively high only in the early part of the nineteenth century. That appears to be the view of both ABM (2002, p. 792–793) and Sokoloff (1986, pp. 686–687, 725, note 4). The implication for the present study is that attention to this issue is required especially for estimating ADO in 1820 and 1831.

However, the problem for these years is the opposite: *underestimation* of ADE, due to absence of data on ADO. Recall that the $AAE(SV)_{AM,NE}$ figures are adopted to develop benchmarks in 1820 and 1831, with 1849 for overlap purpose. SV lack data on days of operation. The main problem is their 1820 sample, which consists directly of annual wage-rate reports. Absent ADO adjustment, the presence of part-time firms downwardly biases $AAE(SV)_{AM,NE}$. SV (1992, p. 37) deal with the problem by removing from their 1820 sample establishments likely to be part-time: "As for the 1820 sample, the bottom 30 percent of the establishments in these [the 18 adopted] industries with the relevant information were truncated from the subsample over which the estimates were prepared[,] to control for the likelihood that a number of firms in 1820 were operating only part of the year and would thus lead to understatement of the annual wage rates prevailing at the time." SV do not state the truncation criterion; but the logical interpretation, provided by Margo (2006c), is that the bottom 30 percent is as measured by the average annual wage.

Fortunately, the problem is essentially absent from the 1831 sample, for two reasons. First, the predominant wage-rate frequency in the McLane Report is daily. Second, the bulk of the establishments in the Report are full-time: "enumerators from the *McLane Report* indicated that nearly all of the establishments covered from the states considered here were operating throughout the year" (Sokoloff, 1986, p. 687—see also ABM, 2002, p. 792, note 1). For 1849, wage data are monthly, which somewhat ameliorates the problem. Also, by that time it may be assumed that full-time operation was the norm, at least for the 18 industries selected by SV. The authors help by eliminating "a small number of outliers" from the 1831 and 1849 [and 1859] samples (SV, 1992, p. 37).

The authors provide evidence that their corrections for part-time firms are sufficient (SV, 1992, p. 39, esp. note 21). Therefore they convert daily and monthly wage rates to annual frequency simply "by assuming 310 days or twelve months of work per year, so as to approximate average annual earnings for full-time employees" (SV, 1992, p. 35). Therefore it is only logical to determine SV average daily earnings from SV average annual earnings as $ADE(SV)_{AM,NE} = AAE(SV)_{AM,NE}/310$, for 1820, 1831, 1849. The SV assumption of 310 days of full-time employment is consistent with the Atack-Bateman (1995, p. 2) and ABM (2002, p. 800; 2003, p. 179, note 22) specification of 309 days. However, as far as the full AHE series is concerned, the ADO figure is irrelevant as long as it is invariant over 1820, 1831, and 1849. The reason is that "adjusted $ADE(SV)_{AM,NE}$" is to be linked to ADE 1849–1919 via the 1849 overlap.

$ADE(SV)_{AM,NE}$ must be adjusted from adult-male, Northeast to all-worker, national coverage. The adjustment is done in two stages. *First,* conversion is to all-worker, retaining the Northeast geographic restriction. Letting $ADE(SV)_{NE}$ denote average daily earnings over all workers in the Northeast, the estimating equation emanates from the definition of $ADE(SV)_{NE}$ in terms of component age-gender wages for which information exists:

$$ADE(SV)_{NE} = E_{AM} \cdot W_{AM} + E_B \cdot W_B + E_F \cdot W_F \qquad (2)$$

where (E_{AM}, E_B, E_F) denotes (adult males, boys, females) as a proportion of all employed workers in manufacturing in the Northeast, and (W_{AM}, W_B, W_F) the (adult-male, boy, female) average daily earnings in manufacturing in the Northeast. Of course, $W_{AM} = ADE(SV)_{AM,NE}$ and $E_{AM} + E_B + E_F = 1$. Dividing by W_{AM} and rearranging terms,

$$ADE(SV)_{NE} = [E_{AM} + E_B \cdot (W_B/W_{AM}) + E_F \cdot (W_F/W_{AM})] \cdot ADE(SV)_{AM,NE} \qquad (3)$$

Assume that relative boy and female (with respect to adult-male) ADE are equal to the corresponding relative average wage. Then W_B/W_{AM} and W_F/W_{AM} are the boy/adult-male and female/adult-male relative-wage ratios. Equation (3) is to be applied to 1820, 1831, and 1849; so values for these years are needed for all right-hand-side variables. $ADE(SV)_{AM,NE}$ is known; it is to be transformed. Estimates of E_{AM}, E_B, E_F are in appendix, *Gender (1800–1859)*. What is required here for the values of W_B/W_{AM} and W_F/W_{AM} are not the "best estimate" of *actual* relative wages but rather, wherever applicable, the relative wages *assumed* by SV to obtain their estimates of $AAE_{AM,NE}$. For 1820, SV (1992, pp. 36–37) specify $W_B/W_{AM} = W_F/W_{AM} = 0.35$; and, for 1849, $W_B/W_{AM} = 0.50$. For the remainder, use is made of the estimates of Goldin and Sokoloff (1982, pp. 760–761; 1984, p. 477), in particular, the average of their figures for New England and the Middle Atlantic. Thus W_B/W_{AM} for 1831 is taken as 0.4005, W_F/W_{AM} for 1831 0.416 and for 1849 0.4805 (see table 5.12 and *Interpolator and Extrapolator Series*). The Goldin-Sokoloff "(a)" figures are selected, because they are based entirely on McLane Report and Census data.

Strictly speaking, one should weight the Goldin-Sokoloff New England and Middle Atlantic figures by the share of boy or female manufacturing workers in the particular year. These data are lacking. However, equal weighting is not an unreasonable approximation, and evidence exists for females. There is a higher female population in the Middle Atlantic throughout 1820–1850. (For the female population in New England and Middle Atlantic states in the Census years 1820, 1830, 1840, 1850, 1860, see Haines, 2006.) However, there are two countervailing elements. The first is the concentration of the female-worker-intensive textile industry in New England.

The second countering factor is higher female labor-participation rates in New England. Estimates of adult (age-16-and-over) female labor-participation rates for 1820, 1830, 1840, 1850, 1860 are provided in Weiss (1992, p. 48). In each year, female participation rates are uniformly higher in each of four New England states (Connecticut,

Massachusetts, New Hampshire, Rhode Island) than in each Middle Atlantic state (New Jersey, New York, Pennsylvania). Also, Goldin and Sokoloff (1982, p. 768) present a table which shows that the 10–29 (or 15–29)-year-old female manufacturing-employment participation rate for New York is less than half that of each of the aforementioned New England states in 1850.

The *second* stage in the conversion process is transforming $ADE(SV)_{NE}$ from Northeast to all-U.S. coverage. This itself is a five-step process.

Step 1: Compute AAE_{NE} for 1849, where AAE_{NE} = average annual earnings, manufacturing, Northeast, Census data. Details of the computation are shown in table 5.9. The computation is on the same basis as for the 1849 AAE for the entire country (see *Average Annual Earnings*, especially table 5.5). All figures in table 5.9 are sums of corresponding individual-state figures.

Table 5.9 Computation of average annual earnings, manufacturing, Northeast[a], 1849—Census data

Item	Wage-earners[b] (number)	Wages[c] (dollars)
All industries[d]: Census	696,661	170,905,581
Deductions		
nonmanufacturing sectors		
fisheries	5552	1,532,292
forestry	123	30,636
mining	13,762	3,767,410
quarrying	73	21,528
construction	10,649	3,946,960
hand and custom trades		
bakers	4600	1,313,592
blacksmiths	13,097	3,732,583
bleachers and dyers	565	161,688
riggers	57	31,464
wheelwrights	6999	1,961,544
All industries: adjusted	641,184	154,405,884
Average annual earnings[e] (dollars)	241	

Notes:
[a] New England and Middle Atlantic states.
[b] Sum of "male hands" and "female hands" in source.
[c] Termed "cost of labor" in source.
[d] Termed "aggregate of manufactures" in source.
[e] Ratio of wages to wage-earners.

Source: All data from Kennedy (1859, pp. 37–136, 143).

Step 2: Obtain the rest-of-U.S./Northeast wage ratio for 1820, 1831, 1849. Consider the equation

$$AAE = E_{NE} \cdot AAE_{NE} + E_R \cdot AAE_R \qquad (4)$$

where (AAE, AAE_{NE}, AAE_R) = average annual earnings, manufacturing (United States, Northeast, rest-of-U.S.), Census data; and (E_{NE}, E_R) = (Northeast, rest-of-U.S.) proportion of U.S. manufacturing employment. $E_{NE} + E_R = 1$. Rearranging terms in equation (4),

$$AAE_R = AAE/E_R - (E_{NE}/E_R) \cdot AAE_{NE} \qquad (5)$$

For 1849, AAE is from table 5.5 and AAE_{NE} from table 5.9 (step 1). Note that the original (unlinked to 1889–1919) AAE is used, because AAE_{NE} is inherently unlinked. For E_{NE} and E_R, 1849 figures are taken from appendix, *Region: Census (1820–1859)*. The resulting AAE_R is $225, compared to $241 for AAE_{NE}.

Let RAAE = ratio of rest-of-U.S. to Northeast average annual earnings, manufacturing, Census data. For 1849, RAAE = AAE_R/AAE_{NE} = 225/241 = 0.93. For t = 1820, 1831, $RAAE_t$ is calculated via the formula

$$RAAE_t = (RAAE_{1849}/RMAR_{1849}) \cdot RMAR_t \qquad (6)$$

where RMAR = ratio of rest-of-US. to Northeast average daily wage, Margo data, from *Interpolator and Extrapolator Series*. The Margo data are for males; so one must assume that the regional relative wage rate for males is applicable to males and females together.

Step 3: Derive the U.S./Northeast wage ratio for 1820, 1831, 1849. The pertinent equation is

$$W = E_{NE} \cdot W_{NE} + E_R \cdot W_R \qquad (7)$$

where (W, W_{NE}, W_R) = (U.S., Northeast, rest-of-U.S.) wage. Dividing by W_{NE},

$$RWUN = E_{NE} + E_R \cdot RAAE \qquad (8)$$

where RWUN = (W/W_{NE}), the U.S./Northeast wage ratio, and W_R/W_{NE} is represented by RAAE. Again, E_{NE} and E_R are in appendix, *Region: Census (1820–1859)*. RWUN is computed for the three years via formula (8).

Step 4: Convert $ADE(SV)_{NE}$ from Northeast to all-U.S. coverage for 1820, 1831, 1849. The formula is

$$ADE(SV) = RWUN \cdot ADE(SV)_{NE} \tag{9}$$

where $ADE(SV)$ = average daily earnings, United States, Sokoloff-Villaflor data

Step 5: Transform $ADE(SV)_t$ to a Census basis (the desired ADE_t) for t =1820, 1831. The formula is

$$ADE_t = [ADE_{1849}/ADE(SV)_{1849}] \cdot ADE(SV)_t \tag{10}$$

Average Daily Hours

Benchmark years for average daily hours (ADH) coincide with benchmark years for ADE only for 1899–1919. For the earlier period, the benchmark years for ADH are fewer and are identical only in one year. Although there exists ADH figures for some other ADE benchmark years, the figures are of a lower level of reliability than the adopted ADH benchmark figures.

All the ADH benchmark figures are "full-time" rather than "actual-work" hours. There is a certain logical consistency with ADE based on days of operation. As Rees (1961, p. 28) comments: "To the extent that it [short-time] takes the form of not working for full days, it is caught in our data on the average number of days in operation per year." However, Rees (and the present author) admit: "To the extent that short-time occurs within days, our method fails to take it into account." Then AHE would be underestimated. Fortunately, Rees finds that, in practice, AHE emanating from full-time rather than actual-work hours is not biased in either direction. In any event, virtually all pre-1914 hours data are full-time.

The Rees ADH series (discussed in chapter 3, COMPOSITE SERIES) provides the benchmark figures for 1890, 1899, 1904, 1909, 1914, and 1919. There is no benchmark figure for 1889. So the 1890 Rees figure (at 10.02 hours) is an ADH benchmark out of sync with the ADE benchmark years. For outline and discussion of the Rees ADH series, see chapter 3, COMPOSITE SERIES; and also Rees (1960, pp. 19–20; 1961, pp. 36–37), Bureau of the Census (1975, p. 155), and Margo (2006a, p. 2.269).

Rees is a clear choice over alternatives. The ADH developed by Brissenden (1929, p. 354), Douglas (1930, pp. 112, 114, 116), and Kendrick (1961, p. 445) are obviously superseded by Rees. Jones' (1963, p. 375) 1900–1919 series is a simple transformation of the

Rees ADH for a crude estimate of average actual hours per week. What is left is the Wolman (1938, p. 2) series, part of which is utilized by Rees.

Another ADH benchmark figure is derived from ABM (2002, p. 798, note 19) data. Using the same 1880 Census sample as for ADO, ABM compute 10.03 and 10.06 as employment-weighted ADH for part-time and full-time establishments. With 59.8 percent of establishments full-time, the weighted-average ADH for all establishments is 10.05. Because of the nature of the Census query (May to November, November to May), it is reasonable to consider the figure as pertaining jointly to 1879 and 1880. That is also the interpretation of Whaples (1990, p. 26).

The final benchmark year is in synchronization with ADE. Based on the Sokoloff sample from the McLane Report, Atack and Bateman (1992, p. 136) calculate an ADH of "about 11 hours 20 minutes a day." So the benchmark figure is 11.33 hours for 1831. To obtain ADH for the benchmark years 1849–1889 and 1820, interpolation and extrapolation of the ADH benchmark figures are necessary. This is performed in appendix, HOURS (1800–1890).

Average hourly earnings (AHE_S) for the benchmark years of the unlinked 1800–1919 segment of the AHE series is readily computed as ADE/ADH. Thus one has AHE_S for 1820, 1831, 1849,..., 1919.

Interpolator and Extrapolator Series

Candidate series for interpolating between AHE_S benchmark figures spanning the period <u>1889–1919</u> are three Douglas series (all, payroll, and union manufacturing industries—shown in table 3.1 and discussed in chapter 3, AVERAGE HOURLY EARNINGS OR HOURLY WAGE RATE) and two Rees series (all-manufacturing and combined-industry, shown in table 3.3 and discussed in chapter 3, COMPOSITE SERIES). All five series have the, desirable, AHE (rather than, say, ADE) characteristic. All have the disadvantage of missing the 1889 observation. The Rees combined-industry series has serious geographic and industry omissions and is constructed only for 1890–1914. This series is very close to the Rees all-manufacturing AHE and is generated by Rees only as a check on that series. Therefore the combined-industry series is dominated by the Rees all-manufacturing AHE. On the other side, the Douglas union series has substantial problems of its own, which carries over into the Douglas all-manufacturing series. What is left is the Douglas payroll series, and Rees himself is responsible for its resurrection: "the defects of the [Douglas] payroll data, though they seemed to be serious a priori, turned out to be surprisingly unimportant in practice" (Rees, 1961, p. 23).

If the Rees all-manufacturing AHE is selected as the interpolator, then one essentially adopts the entire Rees composite series for 1890–1919. Why not do that? The reason is that Douglas payroll is the superior series. First, the Douglas payroll series covers a higher proportion of wage-earners than the Rees AAE interpolator series (an important ingredient in the Rees all-manufacturing AHE), as shown in table 5.10. Second, the Rees series is geographically restricted to three states, whereas the Douglas series is national in scope. Third, while the Douglas series is obtained directly, the Rees series is computed indirectly, using the same methodology as for the benchmark years. It is arguable that, other things being equal, the scope for error is greater under indirect computation. Of course, the Douglas payroll average daily earnings series (DAHE) must be extended to 1889, performed below.

Potential interpolator series for the 1859–1889 period are shown in table 3.2 and evaluated in chapter 3, AVERAGE DAILY WAGE RATE. They have the common denomination ADW rather than, the desired, AHE. Falkner, Mitchell, and Phelps Brown and Hopkins have their own specific problems and in any event are superseded by the various Long series. Long-Weeks does not extend beyond 1880 and is unweighted at firm and state levels. Long-*Bulletin 18* is questionably a manufacturing series and exists only for 1870–1890.

What remains is Long-Aldrich. The series is not used as is. Rather, a revised Long-Aldrich average daily wage series (RLADW) is created.

The revised Long-Aldrich series incorporates three improvements. Component industries of the revised series are the following 11 Long-Aldrich industries: agriculture implements; ale, beer, porter; books and newspapers; carriages and wagons; cotton manufactures; illuminating

Table 5.10 Rees interpolator series and Douglas payroll series: Comparison of coverage

Year	Percent of Wage-Earners	
	Rees	Douglas
1904	16	30
1914	26	28

Source: Rees—Table 3.4. Douglas—computation made by present author, based on data in Douglas (1930, pp. 94, 219).

gas; leather; metals and metallic goods; paper; white lead; woolen manufactures. So one improvement is the omission of nonmanufacturing industries, reducing Long's 13-industry set to 11. For each retained industry, for 1860–1890, the annual ADW is computed as the average of the Long (1960, pp. 121–124) January and July figures. Thus a second improvement is use of both intra-annual figures for an annual series.

The third improvement is the weighting pattern. Long uses Census industry-employment weights, and linearly interpolates between Censuses. There are several problems. (1) The years 1860, 1870, 1880, 1890 are applied in lieu of the correct Census years 1859, 1869, 1879, 1889. (2) Long provides neither the weights themselves, even for Census years, nor the Census source of weights. One can make inferences for some industries from Long's (1960, p. 150) Table A-9. However, it remains unclear which Census industry data are taken for certain industries, in particular "metals and metallic goods" and "white lead." (3) There are inevitable issues of judgment in deriving the employment figures, which are not discussed in Long. The present study uses Census industry-employment weights in a way that avoids these problems. The weights are generated in appendix, *Industry*. Of course, the weighing pattern is proportional to employment; that is, the weights sum to unity—the practice followed for all weighting patterns in the study. The resulting employment-weighted ADW of the 11 stipulated manufacturing industries, annual for 1860–1890, is the RLADW.

The next task is to extend RLADW backward to the year 1859. A reconstructed Falkner wage index is used for this purpose. Consider the 11 industries underlying RLADW. For seven of these industries, from 1860 to 1861 the Falkner component wage index and the RLADW component wage series both increase (agricultural implements, leather, metals and metallic goods, white lead, and woolen manufactures), both decrease (books and newspapers), or both remain unchanged (carriages and wagons). Again using Census industry-employment weights from appendix, *Industry*, the seven-industry average of the Falkner component wage indexes constitutes the reconstructed Falkner average daily wage index (FADW) and is computed for the three years 1859–1861.

$RLADW_{1859}$ is then estimated as $(FADW_{1859}/FADW_{1860}) \cdot RLADW_{1860}$. The most serious objection to this procedure is that the Falkner wage indexes weigh occupations equally irrespective of employment, in contrast to Long's adoption of employment weighting. However, selection of industries with Falkner index positively correlated with the analogous Long series alleviates at least part of the problem. A second issue is that the use of the Falkner industry series destroys Long's 1860–1890 continuity of firms; but Falkner's inclusion of more firms might counteract that loss.

FADW is used to extrapolate RLADW for only one year; longer extrapolation would be questionable. Also, one would want to extrapolate not far into the Civil War period, preferably only backward into the antebellum period. For 1860–1861, movements of the original Long and Falkner series differ by only 7/10ths of one percent (see chapter 3, AVERAGE DAILY WAGE RATE). Considering the revised series, the 1860/1861 ratio of RLADW is 0.975 and of FADW 0.977—an amazingly close correspondence, which supports the estimation procedure for 1859.

The Long series, and therefore the 1859–1890 RLADW, is restricted to the Northeast, that is, New England and the Middle Atlantic regions. (The only anomaly is one Ohio establishment in one industry.) Long (1960, p. 75) includes Delaware and Maryland (and, implicitly, the District of Columbia [DC]) in the Middle Atlantic. Interestingly, that corresponds to the original Census definition of "the Middle States."

However, the employment weights underlying RLADW are national; they are not geographically restricted. So multiplication of RLADW by an adjustment ratio (AR, U.S.-wage/Northeast-wage) estimates a RLADW extended from the Northeast to the United States. The primary source for the data to construct the adjustment-ratio series is Coelho and Shepherd [CS] (1976), but the series cover only 1851–1880. Data in *Bulletin 18* (1898) are used for 1880–1890. For discussion of the CS series, one may consult CS (1976, pp. 206–212, 224–230) and Margo (2000b, pp. 8–9; 2006f).

CS use occupational wage data from the Weeks Report (see chapter 2, *Special Reports* under EARNINGS AND WAGES) to generate aggregate ADW series (and individual-occupation ADW series—not needed here) for the United States and nine regions separately, for 1851–1880. Of the six underlying occupations, only four (engineers, blacksmiths, machinists, common laborers) can be construed as pertaining to manufacturing, even in part, with two (painters, carpenters) clearly belonging to construction. Further, blacksmithing, generally considered a hand trade, is excluded from manufacturing AAE (see tables 5.3, 5.5, and 5.6). However, Atack and Bateman (1999, p. 187) take a different view, noting that "throughout most of the nineteenth century, they [blacksmiths] produced a wide range of goods that fully deserve to be called 'manufactured products.'"

While it might be inappropriate to use the CS *level aggregate* series for the United States or a given region to represent the manufacturing wage for the United States or that region, geographic *ratios* of the six-occupation aggregate series reflect primarily *regional* (or, as in this case, national/regional) *differences* in wages and are thus relatively invariant to industry composition of the series. Therefore the six-occupation adjustment ratio reasonably represents this ratio for the manufacturing sector.

There is also the weighting issue. The number of workers in each occupation-region category is unknown, because the Weeks occupation-establishment-region observation has unknown employment. Therefore, of necessity, the number of observations underlying each CS individual occupation-region wage series (for a given year) consists only of the number of establishments reporting a wage figure for that occupation in that region. This property carries over into the national series. CS provide the sample size (number of observations) that underlies each of their average-wage series in each year. Considering aggregate series, for the year (1859, 1870, 1880) there are (213, 740, 1256) observations for the United States and (125, 427, 613) observations for the Northeast. Note that a particular reporting establishment in a given year can provide more than one observation for an aggregate series, because the establishment may be employing more than one type of worker (out of the six occupation categories).

Again it is arguable that the unweighted nature of individual observations adversely affects the Coelho-Shepherd data only for wage *levels*. For regional, or national/regional, wage *ratios*, the series achieve representativeness. As CS (1976, p. 211) themselves state: "Because these occupational wage averages are unweighted by the proportion of wage earners in each occupation, they cannot be taken to represent average levels of wages in each region. However, they can serve to indicate differences in wage levels among regions."

It is pleasing consistency that the CS composition of the Middle Atlantic region (and, of course, New England) is identical to that of Long. However, these regions' six-occupation wage series must be combined to obtain the Northeast wage. This is readily done by weighting the two regional series, year-by-year 1859–1880, by their respective shares of Northeast observations. CS provide the U.S. aggregate wage series directly. All data are in Coelho and Shepherd (1976, p. 226). Then one readily constructs the CS U.S./Northeast wage-ratio series, constituting the adjustment ratio (ARCS) for 1859–1880.

The regional quality of the CS data is a concern. According to CS (1976, p. 207), the Northeast region (comprised of New England and the Middle Atlantic) has a "large sample size": (125, 427, 613) observations in (1859, 1870, 1880). They warn the reader that, in effect, only one of the seven regions constituting the rest of the United States has a "large sample size." In fact, for three of these regions (West South Central, Mountain, Pacific) "the sample sizes are not large enough to make reliable statements concerning the levels of wages in these regions." Further, six of the seven rest-of-U.S. regions (all except East North Central) in the 1850s "have sample sizes too small to enable one to place much reliance upon wage estimates."

Fortunately, the rest-of-U.S. region (combining seven CS regions) does have a large, or at least moderate, sample size: (88, 313, 643) observations in (1859, 1870, 1880). (All sample-size data are in CS, 1976, pp. 226–227.) Two concerns do remain. First, in general, for each of the nine regions, the sample size increases more or less steadily over time. As a result, it may be that the constructed ratio is less reliable as one moves into the past. Second, considering the rest-of-U.S. composed of the three conventional regions Midwest, South, West; the composition of the rest-of-U.S. sample is far from uniform geographically: the percent of total rest-of-U.S. observations in (1859, 1870, 1880) is (78, 73, 72) for the Midwest, (22, 22, 24) for the South, and (0, 5, 4) for the West. True, Maryland, Delaware, and DC are placed by CS (and Long) in the Northeast rather than the South, which reduces the South absolute number and share of observations. Nevertheless, the nonuniform geographical coverage of the numerator of the adjustment ratio could lead to a biased ratio series.

For 1880–1890, *Bulletin 18* (discussed in chapter 2, *Wages*) is the only source for annual regional wage rates. Long (1960, pp. 86–87) and Rosenbloom (1990, pp. 90–94) use *Bulletin 18* for regional wage comparisons. Both authors have the Long composition of Northeast cities (Baltimore, Boston, New York, Philadelphia, Pittsburgh), and both compute regional ratios with respect to the Northeast. Long shows the median of his ten "manufacturing"-occupation ratios for selected years (see chapter 2, *Wages*). Rosenbloom estimates relative regional *effects* on regional wage ratios—quite different from the ratios themselves because (in terms of his model) occupation effects and residual effects are excluded. With no weighting of occupations or regions, neither author can use his technique to generate a U.S./ Northeast wage ratio comparable to ARCS. Also, the authors' results are for quinqennial years (Long) or multi-year intervals (Rosenbloom) rather than for all years.

However, Long (1960, pp. 135–140) does provide gainfully employed weighted averages of the ten occupation wages for the United States and Northeast annually (see table 3.2). So one could compute the U.S./Northeast wage ratio for 1880–1890 directly from his series. Long's methodology is followed, but his technique is improved, altered, and extended for this study in various ways.

1. Long converts the *Bulletin 18* greenback-period data from gold dollars back into greenbacks. That is unnecessary here, as same-year wage ratios are independent of the currency unit.

2. Long's list of "manufacturing" occupations is adopted, except that "iron molders" and "pattern makers, iron works" are combined (via averaging their wage series) to create the "iron and steel" occupation. The reason is to facilitate the weighting pattern of the wage series (see appendix, OCCUPATION (1880–1890): CENSUS).

3. Long does not distinguish between skilled and unskilled occupations. Weighting is much improved by making this distinction. Skilled workers are blacksmiths, boiler makers, cabinet makers, compositors, iron-and-steel workers, machinists, and stonecutters. Unskilled workers are "laborers (not specified)" [that is, nonstreet laborers] and "teamsters." Teamsters could be classified as semiskilled rather than unskilled workers, but "there is precedent for combining common laborers and teamsters" (Margo and Villaflor, 1987, p. 878; see also Margo, 2000b, p. 43). The fact that blacksmiths and the other occupations are found not only in manufacturing but also in other sectors does not vitiate their use in ratio form.

4. Long weights each city-wage series by the number of gainfully employed workers in the state in which the city is located. There are several problems with this procedure. First, there is a tremendous overweighting of unskilled workers relative to skilled workers. For example, in 1879 nationally "laborers (not specified)" alone constitute 70 percent of gainful employment in the specified "manufacturing" occupations. Adding teamsters, the percentage is 76 percent [Census Office, 1883; and see appendix, OCCUPATION (1880–1890): CENSUS]. The bias occurs for two reasons: (a) the specified skilled occupations are a small subset of all manufacturing skilled occupations, (b) relative to the skilled occupations, the unskilled occupations are less attached to manufacturing than to the economy at large. Second, it seems arbitrary to let a state's employment be attached to a city within the state. The weight may easily be too high or too low. Third, Long provides neither the correspondence of Census occupation data to the wage series nor the weighting patterns. The judgments

involved in creating the former should be made explicit, as they are in the appendix.

In the present study, two regions are distinguished: Northeast and rest-of-U.S. For a given occupation, the Northeast wage is the unweighted average of the (maximum) five Northeast-cities wages, and the rest-of-U.S. wage the unweighted average of the (maximum) seven rest-of-U.S.-cities wages. There is precedent in that *Bulletin 18* itself provides an unweighted average wage of the 12 cities for each occupation. Within the skilled and unskilled categories, for each of the two regions, weights are proportional to Census occupational employment (linearly interpolated for intercensal years), described in appendix, OCCUPATION (1880–1890): CENSUS. Thus a (skilled, unskilled) wage for the Northeast (WSNE, WUNE) and rest-of-U.S. (WSR, WUR) is obtained annually for 1880–1890.

The *Bulletin 18* Northeast wage (WNEB) is the weighted average of WSNE and WUNE, the rest-of-U.S. wage (WRB) the weighted average of WSR and WUR. Weights are derived from the CS (1976) data on sample sizes for the occupations in the Weeks Report. The weights are for 1880—the latest year in the Weeks Report, also the year of maximum sample size. Letting the CS "common laborers" represent all unskilled workers ("laborers (not specified)" and "teamsters"), the weights are (.6705, .3295) for (WSNE, WUNE) and (.6283, .3717) for (WSR, WUR).

There are two problems with these weights. First, the Weeks and CS data pertain to the combined manufacturing, construction, and mining sectors rather than to manufacturing alone (see chapter 2, *Special Reports* under EARNINGS AND WAGES). Still, the adopted weights are much closer to "true" manufacturing weights than the Census employment figures would provide. Second, observations are firms rather than workers within firms. This is a serious limitation; but the adopted weights are superior to (0.5, 0.5) weights—an unweighted average. With "no occupational weights" attached to the Weeks wage series, Lebergott (1964, p. 296) warns against "simple [that is, unweighted] averaging of such wage reports for an establishment, or for an industry." However, the average here is for higher aggregation, "all" manufacturing (plus mining and construction), for which the danger is not as severe. It is also pleasing that the weights using 1879 would be virtually identical to those for 1880.

The *Bulletin 18* U.S. wage (WUSB) is the weighted average of WNEB and WRB. Weights are proportional to Census occupational employment in the respective regions, with linear interpolation for intercensal years [see appendix, OCCUPATION (1880–1890): CENSUS]. The *Bulletin 18* adjustment ratio (ARB) is WUSB/WNEB, computed for 1880–1890. The 1859–1890 adjustment ratio (AR) for RLADW, the revised Long-Aldrich average-daily-wage series, is the CS ratio (ARCS) directly computed for 1859–1880 and estimated as $(ARCS_{1880}/ARB_{1880}) \cdot ARB_t$ for t = 1881–1890. For 1859–1890, ALADW, the RLADW series adjusted from the Northeast to the entire country, is ARCS·RLADW. The adjustment factor ARCS is shown in the second column in table 5.11.

Table 5.11 Ratios to extend coverage of new series

Year	Ratio		
	U.S. Wage to Northeast Wage		Northeast
	Coelho-Shepherd and Bulletin 18 Data (ARCS)	Margo Data (WUS/WNE)	All-Worker Wage to Male Wage (W/W_M)
1808	—	—	0.99
1809	—	—	0.98
1810	—	—	0.97
1811	—	—	0.96
1812	—	—	0.95
1813	—	—	0.94
1814	—	—	0.93
1815	—	—	0.92
1816	—	—	0.90
1817	—	—	0.89
1818	—	—	0.88
1819	—	—	0.87
1820	—	0.92	0.86
1821	—	1.06	0.85
1822	—	1.02	0.85
1823	—	0.99	0.84
1824	—	1.01	0.84
1825	—	1.02	0.84
1826	—	1.05	0.83
1827	—	1.05	0.83
1828	—	1.06	0.83
1829	—	1.07	0.82
1830	—	1.09	0.82
1831	—	1.10	0.82
1832	—	1.08	0.82
1833	—	1.08	0.82
1834	—	1.04	0.82

Continued

Table 5.11 Continued

Year	Ratio		
	U.S. Wage to Northeast Wage		Northeast
	Coelho-Shepherd and Bulletin 18 Data (ARCS)	Margo Data (WUS/WNE)	All-Worker Wage to Male Wage (W/W_M)
1835	—	1.03	0.82
1836	—	1.02	0.82
1837	—	1.06	0.82
1838	—	1.00	0.82
1839	—	1.02	0.82
1840	—	1.04	0.82
1841	—	1.01	0.83
1842	—	1.03	0.83
1843	—	0.97	0.83
1844	—	0.99	0.84
1845	—	0.96	0.84
1846	—	0.96	0.85
1847	—	0.98	0.85
1848	—	0.98	0.85
1849	—	0.98	0.86
1850	—	0.98	0.86
1851	—	1.02	0.86
1852	—	1.00	0.86
1853	—	1.00	0.86
1854	—	1.00	0.86
1855	—	0.99	0.86
1856	—	0.97	0.86
1857	—	0.99	0.86
1858	—	0.99	0.86
1859	1.02	0.98	0.86
1860	1.01	—	—
1861	1.04	—	—
1862	1.07	—	—
1863	1.09	—	—
1864	1.07	—	—
1865	1.04	—	—
1866	1.02	—	—
1867	1.01	—	—
1868	1.00	—	—
1869	1.00	—	—
1870	0.99	—	—
1871	0.99	—	—
1872	0.98	—	—
1873	0.99	—	—
1874	0.99	—	—

Continued

Table 5.11 Continued

Year	Ratio		Northeast
	U.S. Wage to Northeast Wage		
	Coelho-Shepherd and Bulletin 18 Data (ARCS)	Margo Data (WUS/WNE)	All-Worker Wage to Male Wage (W/W_M)
1875	0.99	—	—
1876	0.99	—	—
1877	1.00	—	—
1878	1.00	—	—
1879	1.01	—	—
1880	1.00	—	—
1881	0.99	—	—
1882	0.98	—	—
1883	0.97	—	—
1884	0.96	—	—
1885	0.97	—	—
1886	0.97	—	—
1887	0.97	—	—
1888	0.97	—	—
1889	0.97	—	—
1890	0.97	—	—

Source: see text.

ALADW is converted from average daily wages to average hourly wages via division by the ADH series, the latter constructed in appendix, HOURS (1800–1890). So, for 1859–1890, the adjusted Long-Aldrich average hourly wage, ALAHW, is ALADW/ADH.

The 1890 value of ALAHW is used only to extrapolate DAHE to 1889 (see <u>1889–1919</u>). This is done via the 1890 overlap ratio of the two series. The formula is $DAHE_{1889} = (DAHE_{1890}/ALAHW_{1890}) \cdot ALAHW_{1889}$. The 1890 overlap ratio is 0.90. One reason for a nonunity ratio is that the Douglas series is based on earnings rather than wages. Another reason is the different industry compositions of the two series.

For the antebellum (<u>1800–1859</u>) period, one must search thoroughly to discover wage series that have four desirable qualities: industry or industries within the manufacturing sector (chapter 1, INDUSTRY VERSUS OCCUPATION DATA), inclusion of both skilled and unskilled workers (chapter 1, SKILLED VERSUS UNSKILLED WORKERS), availability for a reasonably long period of time prior to 1860 (for enhancement of interpolative/extrapolative usefulness—chapter 1,

SHORT-RUN VERSUS LONG-RUN SERIES), and covering at least one of the years 1800 and 1859 (for same).

The wage information for Massachusetts workers in Wright (1885) includes both skilled and unskilled workers for 1752–1860, but coverage is not continuous and occupations are not manufacturing-based (see chapter 2, *State Labor Bureaus* under EARNINGS AND WAGES). Layer's (1955) ADE series for the cotton-textile industry, 1825–1914, is confined to skilled workers (see chapter 2, *Antebellum Records of Firms*). Smith (1963) provides wage series of skilled and unskilled occupations for 1828–1881 (again see chapter 2, *Antebellum Records of Firms*), but canal maintenance is decidedly not manufacturing. Margo (2000b) constructs wage series for skilled and unskilled workers for 1821–1860; but, again, civilian work for the army is not manufacturing in nature (see chapter 2, *Records of Civilian Employees of U.S. Army*; and below). Adams (1967, 1968) collects wage rates for several skilled occupations and common labor in Philadelphia over 1785–1830, but the industries are construction and shipbuilding (see chapter 2, *Antebellum Records of Firms*).

What is left are the series of two authors, Zabler (1972) and Adams (1982). These series satisfy all four properties and are the inputs to construct the 1800–1859 interpolator/extrapolator series.

For the period 1802–1860, for the Brandywine region, Adams (1982, pp. 904–905) compiles a continuous series of monthly wages for males based primarily on records of the DuPont gunpowder and textile firms and another textile firm (see chapter 2, *Antebellum Records of Firms*). Adams also provides a series for female workers (see below) and for earnings, but the latter only for decadal intervals. The male wage series is reprinted in Margo (2006e), and is described in Adams (1982, pp. 903–905, 916–917) and Margo (2000b, pp. 13–14; 2006e). The series has not been subject to criticism, and appears to be reliable.

Zabler (1972, pp. 112–113) shows monthly wage series, continuously over 1800–1830, separately for what he classifies as six skilled and five unskilled occupations in iron-producing firms in Pennsylvania (see chapter 2, *Antebellum Records of Firms*). In contrast to Adams, the Zabler series are controversial, and have been both vigorously criticized and vigorously defended. One problem is the isolated and rural location of the iron industry at the time. Adams (1973, p. 90) suggests "the distinct possibility that wage rates in the iron industry are not representative of nonagricultural wage rates in general." He provides evidence that the iron industry was low-wage and the Atlantic region was also low-wage (Adams, 1973, pp. 93–94). Zabler (1973, pp. 95–96) responds that workers nevertheless were drawn from urban areas and that iron was one of the few "true" manufacturing industries at the time.

A second issue is Zabler's classification of "filler" as an unskilled occupation, in spite of the fact that that it is high-paid compared to several skilled occupations (Grosse, 1982, p. 414; Lindert and Williamson, 1982, p. 420). Given particular demand-and-supply configurations, an unskilled occupation on occasion could have higher wages than certain skilled occupations. However, the generally high wage for "filler" does suggest that Grosse and Lindert-Williamson are correct in their criticism. Their same argument applies to "woodcutter." Not only does "woodcutter" have a relatively high wage but also the nature of the work indicates classification as a skilled occupation.

A third problem is that, more generally, the skilled/unskilled relative wage based on Zabler's data is low (Adams, 1973; Margo, 2000b, p. 11). Several explanations have been offered. Zabler (1972, p. 115) argues that there was a relatively low supply of unskilled labor combined with agriculture serving as attractive alternative employment. Adams (1973, pp. 92–93) suggests nonmonetary supplements for carpenters (presumably extendible to skilled workers generally); and Margo, more doubtfully, security of artisan (skilled-worker) employment. However, Williamson (1975, pp. 4–11) documents the consistency of Zabler's data with other skilled/unskilled wage-rate information. So the Zabler series are deemed acceptable for use.

Neither Adams nor Zabler provides employment figures associated with their wage data. To create a Zabler overall wage series, his individual-occupation series are combined as follows. First, one computes separate unweighted averages of the revised seven skilled-occupation series (keeper, carpenter, smith, miller, collier, filler, woodcutter) and the three remaining unskilled series (laborer, teamster, banksman). The series for "clerk," a white-collar occupation, is disregarded.

Second, the resulting skilled and unskilled wage series are averaged using weights derived from CS (1976, pp. 226–230). The weights are the ten-year, 1851–1860, average proportion of skilled-worker observations (0.6736) and common-labor observations (0.3264) in the Northeast. This weighting pattern is subject to the same limitations as the CS weights used for the *Bulletin 18* wage series in 1859–1889. The same justifications also apply. The weights have a firmer basis than the (0.5, 0.5) alternative; and, pleasingly, if annual data over 1851–1860 are considered, the weights have a narrow range: 66–68 percent for skilled occupations, 32–34 percent for unskilled.

The Adams (WA) and constructed Zabler (WZ) wage series share two limitations. First, they pertain only to male workers (for Adams, by direct construction; for Zabler, by the nature of the occupations). Second, they are geographically restricted to the Northeast. Therefore, recalling that each series shows monthly wages, the series can be legitimately combined, but then the combined series will have to be extended to encompass female workers and the rest of the country.

The combined Adams-Zabler male Northeast wage series (WAZMNE) has three segments. For 1802–1830, the series overlap and a weighted average of the constituent series is taken. The weights are proportional to U.S. employment in textiles (for Adams) and iron (for Zabler), and are derived in appendix, **Industry**. One may object that only two of the three principal firms (Charles I. DuPont; Bancroft, Simpson and Eddystone) underlying the Adam series are textile manufacturers, with the remaining firm (E. I. DuPont) a gunpowder producer. Lack of employment data for gunpowder gives one no choice. Substantively, the textile weight is reasonable, because (a) all three firms are in the same geographic area, and (b) the gunpowder firm and one of the textile firms are affiliated.

For 1831–1859, the Adams wage series is linked to the combined series via the 1830 overlap. Specifically, $WAZMNE_t = (WAZMNE_{1830}/WA_{1830}) \cdot WA_t$, for t = 1831–1859. For 1800–1801 a similar technique is used for the Zabler series: $WAZMNE_t = (WAZMNE_{1802}/WZ_{1802}) \cdot WZ_t$. For the entire period 1800–1859, WAZMNE is converted from dollars per month to dollars per day by division by 26, the average number of working days per month according to Adams (1973, p. 92) and the "standard figure" according to Margo (2000b, pp. 38, 166, note 4). Consistently, Atack and Bateman (1995, p. 5) and ABM (2002, p. 798; 2003, p. 179) use 25.75 days. Pleasingly, SV (1992, p. 36), though implicitly, also assume a 26-day month, as their figure of 310 working days per year divided by twelve yields 25.83 working days per month. However, no matter; because the full AHE series is invariant to any figure constant over 1800–1859 (see below).

The resulting series is divided by average daily hours (ADH); so one has constructed the combined Adams-Zabler average hourly wage series (WAZHMNE) for male workers in the Northeast: WAZHMNE = WAZMNE/ADH.

Certainly it is a stretch to represent all Northeast manufacturing by only one or two subregions and only a few industries and firms, even given that the worker composition is to be extended from males to all workers. It must be remembered that the series, after geographic incorporation of the rest of the U.S., is to be used only for interpolation and extrapolation with reference to more-reliable benchmark figures. As a consequence, the days-per-month factor is irrelevant, providing it is constant and positive.

Retaining the Northeast geographic restriction temporarily, the WAZHMNE series needs to be extended to incorporate females. A possible input is Adams' (1982, pp. 911–912) series of the female/male average-wage ratio (W_F/W_M) for manufacturing in the Brandywine region. Unfortunately, it is obvious that this is an inconsistent series. Most of the female-wage observations are a range, and Adams takes the midpoint (except for 1845, probably a clerical error) as the numerator in the ratio. The fact that a single figure ($8.00 per month) that serves as the observation for five years is also the lower end-point for most ranges suggests varying coverage of the female wage and therefore an inconsistent ratio series. It is not surprising that the ratio series varies widely. Therefore the Adams series is not used and incorporation of female workers is performed as a multi-step procedure. A complication is that all other age-gender wage-ratio information is relative to the adult-male wage rather than all-male wage.

Step 1. Compute the female/adult-male average-wage-ratio series (W_F/W_{AM}) for Northeast manufacturing in 1808–1859. The year 1808 is arguably the first year of significant manufacturing employment of workers other than adult males—see table A.4, note c.

a. Assemble or estimate the wage ratio for benchmark years. These years, delimited by data availability, are 1815, 1820, 1831, 1849, and 1859. W_F/W_{AM} for these years is shown in the second column in table 5.12. In the process of obtaining the W_F/W_{AM} benchmark values, the adult-female/adult-male (W_{AF}/W_{AM}) wage ratio is assembled or estimated for certain years, shown in the third column. One may recognize that the values of W_F/W_{AM} for 1831 and 1849 are those used in construction of benchmark ADE (see *Average Daily Earnings*).

b. Interpolate linearly between benchmark years, to obtain W_F/W_{AM} for the remaining years 1815–1859; for 1808–1814, extrapolate the 1815 figure. This interpolation and extrapolation are justified by the findings of Goldin and Sokoloff (1982, pp. 757–766; 1984, pp. 471–481) and Goldin (1990, pp. 63–67). They summarize: "industrial expansion in the Northeast increased the relative wage of women and children.... As the [Northeast] manufacturing sector expanded [from 1820 to 1850], the wage for females relative to that for adult males rose rapidly, achieving by 1850 a value that was almost 90 percent of its long-run level...Wherever manufacturing and the factory system

Table 5.12 Estimated average wage of females relative to adult males, manufacturing, Northeast[a], 1815–1859

Year	Female Age Group		Source[b]
	All W_F/W_{AM}	Adult W_{AF}/W_{AM}	
1815	0.262[c]	—	—
1820	0.261[d]	0.337[c]	adult: Goldin and Sokoloff (1982, p. 760)
1831	0.416[e,f]	0.457[g]	all: Goldin and Sokoloff (1982, p. 760)
1849	0.4805[e,f]		Goldin and Sokoloff (1982, pp. 760–761)
1859	0.471[h]	0.510[i]	—

Notes:
[a] New England and Middle Atlantic states.
[b] Other than stated in notes.
[c] Non-industrial relative wage of 0.288 in Massachusetts (Goldin and Sokoloff, 1982, p. 760; 1984, p. 472) projected to manufacturing, and to entire Northeast via multiplication by 1820 ratio of Northeast to New England adult-female relative wage (Goldin and Sokoloff, 1982, p. 760, and 1820 entry in table): 0.337/0.371.
[d] Applying the technique in note g to 1820, $\alpha = 11.4/20.3 = 0.562$ (from table A.4), and $W_G/W_{AF} = 0.60$ (repeating the 1831 figure, for lack of other information). Now, $W_F/W_{AF} = (0.562)(0.60) + 0.438 = 0.7752$, whence $W_F/W_{AM} = 0.7752 \cdot (W_{AF}/W_{AM})$. With $W_{AF}/W_{AM} = 0.337$, $W_F/W_{AM} = 0.261$.
[e] Average of average figures for New England and Middle Atlantic.
[f] Stated as 1832 and 1850 in source.
[g] Following Goldin and Sokoloff (1984, p. 480, n. 21), $W_F = \alpha \cdot W_G + (1 - \alpha) \cdot W_{AF}$, where α = proportion of female manufacturing workers who are girls. Now, dividing by W_{AF}, $W_F/W_{AF} = \alpha \cdot (W_G/W_{AF}) + (1-\alpha)$. For 1831, $\alpha = 0.225$ (from Table A.4, note h), and Goldin and Sokoloff (1984, p. 480) estimate that $W_G/W_{AF} = 0.60$. So $W_F/W_{AF} = (0.225)(0.60) + 0.775 = 0.91$. It follows that $W_{AF}/W_{AM} = (W_F/W_{AM})/0.91$. With $W_F/W_{AM} = 0.416$, $W_{AF}/W_{AM} = 0.457$. This figure is consistent with a "bias from including girl wages in the 1832 female average [that is, the bias from representing adult females by all females]... on the order of 10 percent," as computed slightly differently by Goldin and Sokoloff (1984, p. 480).
[h] Applying the technique in note g to 1859, $\alpha = 5.3/28.0 = 0.189$ (from table A.4) and $W_G/W_{AF} = 0.60$ (again repeating the 1831 figure, for lack of other information). Now, $W_F/W_{AF} = (0.189)(0.60) + 0.811 = 0.9244$, whence $W_F/W_{AM} = 0.9244 \cdot (W_{AF}/W_{AM})$. With $W_{AF}/W_{AM} = 0.510$, $W_F/W_{AM} = 0.471$.
[i] For 1885, for the United States, $W_{AF}/W_{AM} = 1.00/1.79 = 0.559$, as computed by Goldin (1990, pp. 61–62), using daily wage rates in Long (1960, p. 146), with Commissioner of Labor (1886) the ultimate source. Applying the U.S. figure to the Northeast and linearly interpolating from 1831, $W_{AF}/W_{AM} = 0.510$.
W_G = girl wage.

evolved, the relative wages of females and of boys rose.... The ratio of female to male earnings was exceptionally low in the northeastern states prior to industrialization but rose quickly wherever manufacturing activity spread" (Goldin and Sokoloff, 1982, p. 761; 1984, pp. 462–463, 478; Goldin, 1990, p. 63).

Step 2. Compute the boy/adult-male average-wage-ratio series (W_B/W_{AM}) for Northeast manufacturing in 1808–1859. There is only one benchmark figure: 0.4005, for 1831 (see *Average Daily Earnings*). For all other years, t, $(W_B/W_{AM})_t$ is estimated as the product of $(W_B/W_{AM})_{1831}/(W_F/W_{AM})_{1831}$ and $(W_F/W_{AM})_t$. The constant, 1831, factor is 0.96. Movement of the boy wage in step with the female wage is consistent with the view of Goldin and Sokoloff (1982, p. 758): "we cannot compute a wage rate for boys in 1850. Nevertheless, the increase in the relative wage of boys between 1815 [in agricultural New England] and 1832 [in industrial New England], as well as various impressionistic evidence, suggest that the relative boy wage continued for a time to rise with the relative female wage."

Step 3. Estimate the male/adult-male average-wage-ratio series (W_M/W_{AM}) for Northeast manufacturing in 1808–1859. Denote the wage of (males, boys) in Northeast manufacturing as (W_M, W_B). Let ($E_{AM,M}$, $E_{B,M}$), derived in appendix, **Gender (1800–1859)**, denote (adult males, boys) as a proportion of employed male workers in Northeast manufacturing. Then $W_M = E_{AM,M} \cdot W_{AM} + E_{B,M} \cdot W_B$, where, of course, $E_{AM,M} + E_{B,M} = 1$. Then, dividing by W_{AM}, $(W_M/W_{AM}) = E_{AM,M} + E_{B,M} \cdot (W_B/W_{AM})$, and (W_M/W_{AM}) is computed as such for all years 1808–1859.

Step 4. Compute the female/male average-wage-ratio series (W_F/W_M) for Northeast manufacturing in 1808–1859. This is done simply via the equation $W_F/W_M = (W_F/W_{AM})/(W_M/W_{AM})$.

Step 5. Finally, compute the all-worker/male wage-ratio series (W/W_M) for Northeast manufacturing in 1808–1859. Let W denote the wage of all workers in Northeast manufacturing. Letting E_M (E_F) denote males (females) as a proportion of all workers in Northeast manufacturing, obtained from appendix, **Gender (1800–1859)**, with $E_M + E_F = 1$, then $W = E_M W_M + E_F \cdot W_F$. Dividing by W_M, $(W/W_M) = E_M + E_F \cdot (W_F/W_M)$. The (W/W_M) series is constructed as such for all years 1808–1859, and is shown in the final column of table 5.11.

Step 6. Estimate the Adams-Zabler all-worker average-hourly-wage series (WAZHNE). For 1808–1859, WAZHNE is computed as the product of (W/W_M) and WAZHMNE. For 1800–1807, WAZHNE = WAZHMNE.

What remains is to extend WAZHNE from the Northeast to the entire country. For this purpose, regional wage series are required. The CS data are rejected. because their series begin only in 1851 and, as CS acknowledge, their sample is small for the 1850s. What remains are the Margo (2000b, Tables 3A.5, 3A.6) regional series (reprinted in Margo, 2006d), revised and improved over the corresponding Margo-Villaflor (1987, pp. 893–894) series.

Margo makes effective use of the records of civilian employees of the U.S. Army and the Census of Social Statistics (see chapter 2: *Special Reports* under EARNINGS AND WAGES, and *Records of Civilian Employees of U.S. Army*) to construct an ADW series separately for unskilled labor and artisans for four regions (Northeast, Midwest, South Central, South Atlantic) for 1821–1860. Disregarded here are the white-collar wage series and the California series; the former for lack of relevance, the latter for Margo's lack of integration into his four-region series. Discussions of Margo's series are in Margo and Villaflor (1987, pp. 875–880), Margo (2000b, pp. 36–49; 2006d; 2006l, p. 2.43), Oaxaca (2000, p. 1155), Rosenbloom (2000), and Whaples (2001b).

Although the number of wage-month-fort-occupation observations is large (almost 47,000—see Margo, 2000b, p. 29), the many gaps in the data categories induce Margo to employ a hedonic-regression procedure for each of the eight series (unskilled, artisan; four regions), with the resulting figures (obtained from coefficients on dichotomous time-period variables) benchmarked to year-1850 wages estimated from the Census of Social Statistics. Though Margo's wage series are an impressive addition to antebellum data, they do have problems, some of which the reviewers and (to a significant extent) Margo himself acknowledge.

First, the skilled occupations, characterized by artisan (mainly construction) trades, are not manufacturing-specific, and unskilled workers (common laborers, teamsters) are of a general nature. This issue does not preclude relative-regional use of Margo's series.

Second, there are problems with Margo's estimation technique. The time-period dummies are not interactive with other explanatory variables (fort location, season, worker and job characteristics, occupation) and sometimes pertain to multiple years. Also, some ad hoc adjustments are made to the hedonic results (Margo, 2000b, p. 46).

Third, the distribution of observations is skewed temporally and geographically. There is relatively low coverage of the 1820s. As Margo (1992, p. 181) acknowledges: "Because the army was charged with forging a path to the frontier, the composition of the *Reports* sample with respect to location, timing, and occupation differs from what a purely random sample of the antebellum population would yield." Thus frontier forts "that are often on the edge of a large census region" (Whaples, 2001b, p. 201) are overrepresented. The great majority of observations for the Midwest are from Kansas; and, for the South Atlantic, from Florida. Where forts are isolated, dangerous, and engaged in combat, their wage payments might not be legitimately extendible to the entire region.

More generally, the issue of whether wages at army forts are representative of wages in the civilian sector is carefully addressed by Margo. The concern is that the Army lacked the incentive to minimize cost of a given task and therefore overpaid its workers. Fortunately, the empirical evidence assembled by Margo (2000b, pp. 32–34) suggests otherwise. "In sum, it would appear that, in terms of compensating its civilian employees, the army simply paid the going wage in the local civilian market" (Margo, 2000b, p. 34). This finding does not obviate Whaple's criticism that the Margo wage series may overly reflect specific frontier and military events.

None of Margo's eight wage series has an observation for 1820, and half the series begin later than 1821. The objective here is to construct a U.S./Northeast wage series from 1820 to 1859, so that adjustment of WAZHNE from the Northeast to the United States can be performed for that period. Some Margo-Villaflor series begin earlier than the corresponding Margo series, and, where evidence of co-movement exists, are used to extend the corresponding Margo series back in time.

Let M_t and MV_t denote Margo and Margo-Villaflor wage series for the same occupation group and same region in year t. The Margo Northeast artisan series is extended from 1821 to 1820 via the equation $M_{1820} = (M_{1821}/MV_{1821}) \cdot MV_{1820}$. Justification is $(M_{1821}/MV_{1821}) = 0.909$, close to $(M_{1822}/MV_{1822}) = 0.906$. The South Central artisan series is extended from 1821 to 1820 via the same equation and justification, with ratios 1.315 and 1.316.

Let A_t and U_t denote Margo artisan and unskilled wage series for the same region in year t. The South Central unskilled series is now extended from 1821 to 1820 via the equation $U_{1820} = (U_{1821}/A_{1821}) \cdot A_{1820}$, with justification $(U_{1821}/A_{1821}) = 0.443$ and $(U_{1822}/A_{1822}) = 0.434$. Let SA_t denote the South Atlantic unskilled wage series and SC_t the South Central unskilled wage series in year t. Then SA_t, for t 1823 and 1824, is estimated via the equation $(SA_{1825}/SC_{1825}) \cdot SC_t$, with justification $(SA_{1825}/SC_{1825}) = 0.865$ and $(SA_{1826}/SC_{1826}) = 0.855$. Unchanged are beginning years of Northeast unskilled (1821), Midwest unskilled (1823), Midwest artisan (1822), and South Atlantic artisan (1823).

To combine artisan and unskilled wage series, the CS 1851–1860 (skilled, unskilled) figures are used as for the Zabler data: (0.6736, 0.3264) for the Northeast, (0.5874, 0.4126) for the rest-of-United States. Then the Northeast wage (WNE) for 1821–1859 is the appropriate weighted average of the artisan and unskilled wages. The series is extended to 1820 via the equation $WNE_{1820} = (WNE_{1821}/WNEA_{1821}) \cdot WNEA_{1820}$, where WNEA denotes the Northeast artisan wage. The ratio factor is 0.928.

To construct the rest-of-U.S. artisan (WRA) and unskilled wage (WRU) series, use is made of the regional-weights series derived in Margo (2000b, pp. 114–115, with discussion on pp. 103–104). For 1823–1859, both the artisan and the unskilled wage series are constructed as the weighted average of the Midwest, South Central, and South Atlantic series, with weights proportional to the Margo regional weights.

The artisan wage series (WRA) extension to 1820 is performed as follows. For 1822–1823, a partial series (WPRA) is constructed as the weighted average of the Midwest and South Central artisan wage. Then

$WRA_{1822} = (WRA_{1823}/WPRA_{1823}) \cdot WPRA_{1822}$, with the factor ratio 0.960. For 1820–1821, observations exist only for the South Central series (WSCA). Therefore, of necessity, for 1820–1821, $WRA_t = (WRA_{1822}/WSCA_{1822}) \cdot WSCA_t$, with the factor ratio 0.834.

The technique to extend the unskilled-wage series (WRU) is simpler, because the Midwest series begins in 1823 rather than 1822. Denote the South Central unskilled wage as WSCU. For 1820–1822, $WRU_t = (WRU_{1823}/WSCU_{1823}) \cdot WSCU_t$, with the factor ratio 0.847. The rest-of-U.S. wage series (WR_t), for 1820–1859, is computed as $0.5874 \cdot WRA_t + 0.4126 \cdot WRU_t$.

The rest-of-U.S./Northeast wage ratio (RMAR, for *Average Daily Earnings*) is obtained as WR/WNE. The U.S. wage (WUS) is the weighted average of WNE and WR, with weights proportional to manufacturing employment in the Northeast and rest-of-U.S. [E_{NE}, E_R—see appendix, ***Region: Census (1820–1859)***]. The U.S./Northeast wage ratio, WUS/WNE, is shown in the third column of table 5.11. Then, for 1820–1859, the Adams-Zabler U.S. average hourly wage (WAZHUS) is the product of (WUS/WNE) and WAZHNE.

Instead of WUS as computed, there is the alternative of combining Margo's (2000b, Table 5B.4) own U.S. (four-region aggregate) artisan and unskilled wage series. This alternative is rejected because: (1) there are missing years (1820–1822 for artisans, 1820–1824 for unskilled workers); (2) figures of greater precision are obtained via computation from first principles; (3) Margo's weights to combine the Northeast and (implicitly) rest-of-U.S. wages are occupational rather than restricted to manufacturing.

Table 5.13 Interpolation of average hourly earnings[a] between benchmark years

Segment	Interpolator Series[b]	Beginning/Ending Pairs of Years[c]
1889–1919	DAHE	1889/1899, 1899/1904, 1904/1909, 1909/1914, 1914/1919
1859–1889	ALAHW	1859/1869, 1869/1879, 1879/1889
1820–1859	WAZHUS	1820/1831, 1831/1849, 1849/1859

Notes:
[a] Unlinked 1800–1919 segment (AHES)—see *Average Daily Hours*.
[b] Denoted as X in the subsection 1920–2006. Desired series (Y) is AHE_S.
[c] Denoted generally as subscripts 0 and n in the subsection 1920–2006.

Interpolation and Extrapolation

To interpolate between benchmark figures, one uses the technique outlined at the beginning of the chapter. Table 5.13 summarizes the interpolation procedures.

AHE_S is now constructed for 1820–1919. To extend AHE_S to 1800, one uses the formula $AHE_{S,t} = (AHE_{S,1820}/WAZHNE_{1820}) \cdot WAZHNE_t$, for t = 1800,..., 1819. The factor ratio is 0.970.

Linking of 1800–1919 to 1920–2006 Segment

Finally, AHE_S (1800–1919) must be connected to AHE (1920–2006). Rees has a problem of the same nature but greater magnitude—a 13-year gap between his 1890–1919 and 1932–1957 segments (see chapter 3, COMPOSITE SERIES). Here the gap is only one year to be crossed. The obvious extrapolating series is DAHE. First, estimate $AHE_{1919} = (AHE_{1920}/DAHE_{1920}) \cdot DAHE_{1919}$, with factor ratio 0.957. Second, estimate $AHE_t = (AHE_{1919}/AHE_{S,1919}) \cdot AHE_{S,t}$, for t = 1800,..., 1918, with factor ratio 0.854.

Thus one has the average-hourly-earnings (AHE) component of average hourly compensation (AHC) for the full period, 1800–2006. In chapter 6 the average-hourly-benefits (AHB) component is derived. All three series are listed in table 7.1.

CHAPTER 6

Average Hourly Benefits

1929–2006

Although by no means perfect, the benefits data situation for 1929–2006 is much superior to that of the pre-1929 period. Recall that that a gross-earnings rather than regular-earnings concept is used for average hourly earnings (AHE), wherefore average hourly benefits (AHB) are residual in nature (see chapter 1, GROSS EARNINGS VERSUS REGULAR EARNINGS). The technique of constructing AHB involves estimating the proportion mark-up of benefits over earnings (PM_p, the ratio of benefits to earnings, with subscript p denoting production workers in manufacturing) and then multiplying AHE by this ratio to obtain AHB, which is benefits per work-hour. Sources of benefits data in manufacturing are discussed in chapter 2, BENEFITS, especially table 2.4. Census data combine production with nonproduction workers and are available only for limited years. Therefore, to construct PM_p, one begins with Employer Costs for Employee Compensation (ECEC) series (expressed as dollars per hour worked); but two adjustments to these series are required. First, quarterly wages and benefits are averaged to obtain annual figures (North American Industry Classification System [NAICS], 2004–2006; Standard Industrial Classification System [SIC], 2002–2003). Second, net earnings and total benefits are converted to gross earnings and residual benefits (all workers: SIC, 1986–1988; NAICS, 2004–2006; production workers: SIC, 1988–2003; NAICS, 4Q 2006–3Q 2007). Conversion formulas are: gross earnings *equal* net earnings *plus* paid leave *plus* supplemental pay; residual benefits *equal* gross benefits *minus* paid leave *minus* supplemental pay. Thus PM_p is computed for all corresponding ECEC series.

ECEC NAICS production-worker data begin in 4Q 2006. To convert the NAICS all-workers 2004–2006 PM from an all-workers to production-workers basis, let subscript a denote all workers in manufacturing. Construct a four-quarter (4Q 2006, 1Q-3Q 2007) average of PM_a and PM_p, represented as PM_a^{4Qav} and PM_p^{4Qav}. Then, for t = 2004–2006, $PM_{p,t} = (PM_p^{4Qav}/PM_a^{4Qav}) \cdot PM_{a,t}$. The four-quarter ratio factor is 1.096. In contrast, the SIC PM_p is directly computed for 1988–2003. However, for 1986–1987, again only all-workers data are available. Now the 1988 ratio serves as the conversion factor, wherefore, for t = 1986–1987, $PM_{p,t} = (PM_{p,1988}/PM_{a,1988}) \cdot PM_{a,t}$. The ratio factor is 1.089, pleasingly close to the 1989 ratio (1.096—coincident figure to the four-quarter ratio factor).

Now one jumps to the beginning of the period. For 1929–1957, PM_p is the ratio of Rees' (1960, pp. 3–4) manufacturing wage-earner series: "wage supplements per hour at work" (AHB) divided by "average earnings per hour at work" (AHE). Rees' AHE series is discussed in chapter 3, COMPOSITE SERIES. His AHB series is described in Rees (1960, pp. 6, 21–26). He obtains the AHB series by estimating the "supplements-to-wages" component of the Bureau of Economic Analysis (BEA) national-accounts "supplements-to-wages-and-salaries" series. Consistency with the present study is assured, because both the Census (data source of the Rees AHE) and the BEA follow a gross-earnings, and therefore net benefits, methodology.

For the 1958–1977 period, benchmark values of PM_p are provided by the Rees 1957 figure and Employer Expenditures for Employee Compensation (EEEC) PM_p—computed directly because of the gross-earnings concept—for 1959, 1962, 1966, 1968, 1970, 1972, 1974, and 1977. Intervening years are obtained via the usual interpolation procedure, with PM_p the desired series and BEA unadjusted PM_a the interpolator series. Even though the BEA data cover both wage-earners and salaried workers, the closeness to the adopted PM_p figures is remarkable. Figures are shown for the EEEC-available years over 1959–1977 in table 6.1 (the 1957 present-study figure is the Rees PM_p). Further, Jones (1961, pp. 37–39; 1963, p. 385) finds that, for 1929–1957, the difference between the Rees wage-earner AHB and the BEA all-worker AHB is minimal—one cent per hour for 1951 and at most one-tenth of a cent for the other years.

Therefore it would appear logical to continue to interpolate the remaining years, 1978–1985, with 1977 and 1986 the end-points and the BEA PM_a the interpolator series. However, as table 6.1 shows, by 1986 the conformity between PM_p and PM_a had vanished, at least for the data sources at hand. During this time, either the manufacturing production-worker benefits mark-up increased substantially more than the manufacturing nonproduction-worker mark-up and/or there was a divergence in benefits coverage between the Bureau of Labor

Table 6.1 Comparison of benefits mark-up over earnings[a], manufacturing: Present study versus Bureau of Economic Analysis, 1957–1988

Year	Percentage Markup		
	Present Study (production workers)	Bureau of Economic Analysis (all workers)	Ratio of Present Study to Bureau of Economic Analysis
1957	8.9	9.4	0.95
1959	9.7	10.2	0.95
1962	11.3	11.5	0.99
1966	12.6	12.7	0.99
1968	13.2	13.3	0.99
1970	14.3	14.2	1.01
1972	16.3	15.9	1.03
1974	18.5	18.0	1.03
1977	21.4	21.4	1.00
1986	26.4	22.3	1.18
1987	26.1	22.5	1.16
1988	27.1	22.4	1.21

Note:
[a] Wages and salaries, for Bureau of Economic Analysis series.
Source: Bureau of Economic Analysis Web site (www.bea.gov). For present study, see text.

Statistics (BLS) and BEA. In either event, the adopted interpolation procedure is not appropriate. Instead, the 1978–1985 segment of PM_p is estimated by linear interpolation between the PM_p for 1977 and 1986.

It is now a simple matter to construct AHB for 1929–2006 as the product of PM_p and AHE.

1900–1928

The U.S. official national accounts begin in 1929; therefore Rees' derivation of AHB also begins in that year. Although some pre-1929 national-accounts series at higher levels of aggregation have been constructed by private scholars (see, e.g., Carter, James, and Sutch, 2006; James and Sylla, 2006; Rhode and Sutch, 2006; and Sutch, 2006c), the series "supplements to wages and salaries" and any components thereof are not among them. Therefore a creative approach is required to estimate pre-1929 AHB.

<u>Components of Benefits:</u> U.S. Department of Commerce (1954, p. 210) provides a breakdown of "supplements to wages and salaries" for 1929, though only at the aggregate level. The only two items that

pertain to manufacturing are "compensation for injuries" and "employer contributions to private pension and welfare funds." The first item is better termed "workers'-compensation benefits" (WCB) and the second "pension-and-welfare benefits" (PWB). Making the assumption that the all-economy proportions of the two components (0.6219, 0.3781) apply to manufacturing, then $AHB_t = WCB_t + PWB_t$, where $WCB_{1929} = 0.6219 \cdot AHB_{1929}$ and $PWB_{1929} = 0.3781 \cdot AHB_{1929}$.

WCB and PWB are expressed per work-hour, and t applies to 1929 and earlier years. The higher share of WCB is consistent with Rees' (1960, p. 6) comment on his 1929 AHB figure, "most of which probably represented the cost of the workmen's compensation." The technique here involves estimating WCB_t and PWB_t for the earlier years via proxy-variable index-numbers emanating from 1929.

To find proxy variables for the index numbers, one must know the composition of each component. For 1929 and later years, U.S. Department of Commerce (1954, pp. 74–75) provides information. For earlier years, the works of private scholars are consulted. In 1929, workers'-compensation benefits consisted of payments to workers (and their dependents or survivors) under workers'-compensation laws. These laws involve no-fault employer-provided compensation for workplace injuries, and establish funding via private insurance carriers and/or a state fund, with the option for employers to drop out with proof of sufficient resources to cover prospective payments under the law. Workers' compensation has the distinction of being the first legally required benefits, and is a state-initiated and state-run program. Thus workers' compensation is based on laws of the individual states, such legislation enacted beginning in 1911. It follows that WCB can be nonzero only from 1911 onward. For the early history of U.S. workers' compensation, the best sources are Fishback and Kantor (1996, 1998, 2000) and Fishback and Thomasson (2006, pp. 2.708–709).

Turning to "pension-and-welfare benefits," note that prior to the Social Security Act of 1935 there were no legally required social-insurance programs; so the constituent items of PWB are voluntary plans and programs. U.S. Department of Commerce (1954) distinguishes three 1929 components, presumably discussed in order of importance: (1) employer contributions to private pension plans, (2) contributions by employers under health and welfare programs, and (3) employer contributions for group insurance. About twice the amount of text is devoted to discussion of (1) than other researchers: "The 1929 figure [for wage supplements] was only 0.4 cents per hour at work...The amount in earlier years must have been smaller still" (Rees, 1960, p. 6); "[In] 1915–29...workers received virtually all of their compensation in the form of wages and salaries" (Wiatrowski, 1990, p. 29). The AHB series, now denominated in *dollars* per work-hour, is shown in table 7.1 for the entire 1900–2006 period.

CHAPTER 7

Nominal Compensation, Real Compensation, and Standard of Living

COMPENSATION AND ITS COMPONENTS

Table 7.1 presents the main results of the book: the time series of average hourly compensation (AHC, the sum of AHE and AHB), average hourly earnings (AHE, constructed in chapter 5), and average hourly benefits (AHB, constructed in chapter 6). AHB is assumed zero until 1900, then computed for positive values but rounds up to a level of one-tenth of one cent only in 1912. The three variables are rounded to a tenth of a cent (that is, shown to three decimal places) until AHB reaches one cent, which happens in 1936. From then on, the variables are rounded to the nearest cent.

There is a tremendous increase in AHC over the two centuries—understandable because all three variables are measured in nominal (money) terms, that is, they incorporate inflation. The growth in compensation is so great that it can be graphed meaningfully only in logarithmic (ratio) scale, done in figure 7.1. Note that equal distances on the vertical axis represent equal percentage (not equal absolute-dollar) increases in AHC.

The composition of AHC is of great interest. The ratio of benefits to compensation, taken as a percent, is 100·(AHB/AHC) and plotted in figure 7.2. The proportion mark-up of benefits over compensation (AHB/AHC) is different from, and smaller than, the proportion mark-up of benefits over earnings (AHB/AHE), which is used to derive AHB for 1929–2006 (chapter 6, 1929–2006). Also, the gross-earnings foundation of AHE and the consequent residual concept of AHB imply a lower benefits/compensation ratio than otherwise (see chapter 1, GROSS EARNINGS VERSUS REGULAR EARNINGS; chapter 4, *Average Hourly Benefits*; and chapter 6, 1929–2006).

Table 7.1 Average hourly compensation, earnings, and benefits: 1800–2006 (dollars per work-hour)

Year	AHC	AHE	AHB	Year	AHC	AHE	AHB
1800	0.040	0.040	0	1845	0.057	0.057	0
1801	0.040	0.040	0	1846	0.057	0.057	0
1802	0.044	0.044	0	1847	0.061	0.061	0
1803	0.044	0.044	0	1848	0.065	0.065	0
1804	0.046	0.046	0	1849	0.063	0.063	0
1805	0.047	0.047	0	1850	0.061	0.061	0
1806	0.046	0.046	0	1851	0.064	0.064	0
1807	0.046	0.046	0	1852	0.067	0.067	0
1808	0.047	0.047	0	1853	0.068	0.068	0
1809	0.048	0.048	0	1854	0.068	0.068	0
1810	0.046	0.046	0	1855	0.068	0.068	0
1811	0.051	0.051	0	1856	0.067	0.067	0
1812	0.052	0.052	0	1857	0.069	0.069	0
1813	0.050	0.050	0	1858	0.075	0.075	0
1814	0.051	0.051	0	1859	0.076	0.076	0
1815	0.051	0.051	0	1860	0.077	0.077	0
1816	0.049	0.049	0	1861	0.081	0.081	0
1817	0.047	0.047	0	1862	0.091	0.091	0
1818	0.047	0.047	0	1863	0.096	0.096	0
1819	0.045	0.045	0	1864	0.105	0.105	0
1820	0.044	0.044	0	1865	0.112	0.112	0
1821	0.050	0.050	0	1866	0.114	0.114	0
1822	0.046	0.046	0	1867	0.112	0.112	0
1823	0.046	0.046	0	1868	0.112	0.112	0
1824	0.049	0.049	0	1869	0.113	0.113	0
1825	0.048	0.048	0	1870	0.113	0.113	0
1826	0.051	0.051	0	1871	0.116	0.116	0
1827	0.050	0.050	0	1872	0.117	0.117	0
1828	0.048	0.048	0	1873	0.120	0.120	0
1829	0.055	0.055	0	1874	0.118	0.118	0
1830	0.057	0.057	0	1875	0.116	0.116	0
1831	0.056	0.056	0	1876	0.114	0.114	0
1832	0.052	0.052	0	1877	0.110	0.110	0
1833	0.057	0.057	0	1878	0.108	0.108	0
1834	0.052	0.052	0	1879	0.107	0.107	0
1835	0.054	0.054	0	1880	0.111	0.111	0
1836	0.052	0.052	0	1881	0.110	0.110	0
1837	0.061	0.061	0	1882	0.113	0.113	0
1838	0.058	0.058	0	1883	0.114	0.114	0
1839	0.058	0.058	0	1884	0.116	0.116	0
1840	0.057	0.057	0	1885	0.116	0.116	0
1841	0.058	0.058	0	1886	0.119	0.119	0
1842	0.064	0.064	0	1887	0.126	0.126	0
1843	0.056	0.056	0	1888	0.128	0.128	0
1844	0.057	0.057	0	1889	0.133	0.133	0

Continued

Table 7.1 Continued

Year	AHC	AHE	AHB	Year	AHC	AHE	AHB
1890	0.133	0.133	0	1936	0.55	0.54	0.01
1891	0.133	0.133	0	1937	0.63	0.61	0.03
1892	0.132	0.132	0	1938	0.64	0.60	0.04
1893	0.135	0.135	0	1939	0.64	0.60	0.04
1894	0.126	0.126	0	1940	0.67	0.63	0.04
1895	0.126	0.126	0	1941	0.74	0.70	0.04
1896	0.128	0.128	0	1942	0.86	0.83	0.04
1897	0.127	0.127	0	1943	0.98	0.93	0.04
1898	0.128	0.128	0	1944	1.05	1.00	0.05
1899	0.131	0.131	0	1945	1.06	1.01	0.05
1900	0.137	0.137	0.000	1946	1.13	1.08	0.05
1901	0.139	0.139	0.000	1947	1.30	1.24	0.06
1902	0.148	0.148	0.000	1948	1.41	1.35	0.06
1903	0.154	0.154	0.000	1949	1.46	1.39	0.07
1904	0.152	0.152	0.000	1950	1.55	1.46	0.09
1905	0.156	0.156	0.000	1951	1.72	1.61	0.11
1906	0.163	0.163	0.000	1952	1.83	1.71	0.12
1907	0.173	0.173	0.000	1953	1.94	1.81	0.13
1908	0.163	0.163	0.000	1954	1.97	1.83	0.14
1909	0.167	0.167	0.000	1955	2.05	1.90	0.15
1910	0.175	0.175	0.000	1956	2.16	1.99	0.16
1911	0.178	0.178	0.000	1957	2.24	2.06	0.18
1912	0.187	0.186	0.001	1958	2.39	2.19	0.20
1913	0.197	0.196	0.001	1959	2.45	2.24	0.22
1914	0.199	0.198	0.001	1960	2.54	2.30	0.24
1915	0.200	0.198	0.002	1961	2.60	2.35	0.25
1916	0.237	0.235	0.002	1962	2.71	2.44	0.28
1917	0.285	0.283	0.002	1963	2.83	2.53	0.29
1918	0.358	0.356	0.002	1964	2.89	2.61	0.29
1919	0.431	0.429	0.002	1965	3.00	2.69	0.32
1920	0.539	0.537	0.003	1966	3.14	2.78	0.35
1921	0.483	0.481	0.003	1967	3.29	2.92	0.37
1922	0.444	0.441	0.003	1968	3.52	3.11	0.41
1923	0.481	0.478	0.003	1969	3.72	3.27	0.45
1924	0.507	0.504	0.003	1970	3.93	3.43	0.49
1925	0.503	0.499	0.004	1971	4.26	3.69	0.57
1926	0.510	0.506	0.004	1972	4.59	3.95	0.64
1927	0.516	0.512	0.004	1973	4.95	4.21	0.74
1928	0.519	0.515	0.004	1974	5.44	4.59	0.85
1929	0.516	0.512	0.004	1975	6.02	5.04	0.98
1930	0.527	0.523	0.004	1976	6.53	5.43	1.11
1931	0.513	0.509	0.004	1977	7.15	5.89	1.26
1932	0.446	0.441	0.005	1978	7.77	6.37	1.40
1933	0.441	0.437	0.004	1979	8.34	6.81	1.53
1934	0.527	0.523	0.004	1980	9.12	7.41	1.71
1935	0.542	0.537	0.005	1981	10.00	8.09	1.91

Continued

Table 7.1 Continued

Year	AHC	AHE	AHB	Year	AHC	AHE	AHB
1982	10.80	8.70	2.10	1995	16.66	12.67	3.99
1983	11.22	9.00	2.22	1996	16.84	12.97	3.86
1984	11.78	9.41	2.38	1997	18.12	13.99	4.13
1985	12.50	9.94	2.56	1998	18.18	14.20	3.99
1986	12.90	10.21	2.69	1999	18.75	14.70	4.05
1987	13.05	10.35	2.70	2000	19.36	15.17	4.19
1988	13.58	10.68	2.90	2001	19.36	15.29	4.07
1989	14.00	10.95	3.04	2002	21.02	16.47	4.55
1990	14.41	11.25	3.16	2003	21.54	16.65	4.90
1991	14.93	11.57	3.36	2004	23.07	17.26	5.81
1992	15.63	11.95	3.68	2005	23.92	17.74	6.19
1993	16.12	12.17	3.95	2006	24.37	18.33	6.05
1994	16.56	12.40	4.16				

Note: AHE and AHB may not sum exactly to AHC, due to rounding.

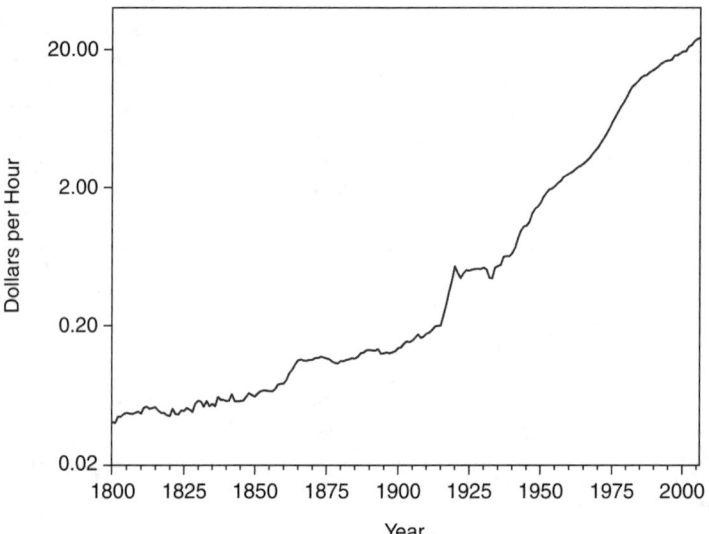

Figure 7.1 Average hourly compensation (logarithmic scale)

As mentioned before, until 1900, AHB is so low that it is taken literally as zero. As the graph shows, while the benefits/compensation ratio has an upward trend, the increase is not steady. Benefits reach one percent of compensation only in 1932, fall below that level for three years; exceed five percent in 1938–1940, but fall below five presented elsewhere by the present author (Officer, 2007a, 2008a). Then real average hourly compensation (AHCR) is constructed as AHC/(CPI/100). AHCR is denominated in "1982–1984 dollars per work-hour," listed in table 7.2, and graphed in figure 7.3.

AHCR increases 37-fold from 1800 to 2006, a far lesser magnitude than for nominal compensation. On the one hand, one sees that in earlier years the standard of living of production workers was greater than a comparison of values of the nominal series over time indicates. On the other hand, any CPI series is beset with problems—such as changes in quality of existing commodities, introduction of new commodities, and omission of important commodities—that tend to bias the series upward as one moves forward in time. So there is a sense in which even AHCR understates improvements in the standard of living over time.

Also, it should be remembered that it is the standard of living of *production workers in manufacturing* that is being measured. The CPI series is based on the official Bureau of Labor Statistics (BLS) series for 1917–2006. Until 1978 the official series relates to urban wage-earners and clerical workers. From that date, the series pertains to all urban consumers. To the extent that the consumption pattern of manufacturing production workers differs from the patterns of

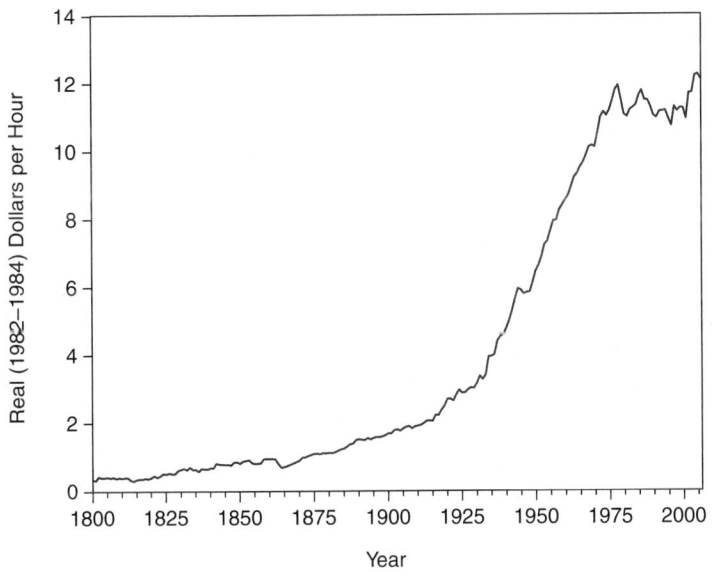

Figure 7.3 Real average hourly compensation

these groups, the AHCR series incorporates conceptual error. Also, the quality of the CPI series generally deteriorates as one goes backward in time—as it usually does for economic data (including the AHC series).

An alternative measure of the standard of living of manufacturing production workers, original to the present study, is the number of work-hours required to purchase "the consumer bundle." The "value of the consumer bundle" (VCB) is a term invented by the present author (Officer, 2007b, 2008b) to describe the "average annual expenditures [per consumer unit]," a BLS series that this author extends back to 1900. VCB emanates from earlier terms—"value of the household bundle" (VHB) and "cost of the (average) household bundle"—developed by Officer and Williamson (2006). VCB is the preferred term, because a "consumer unit" is not the same as a "household." While a household consists of all persons occupying a housing unit, a consumer unit is the decision-making unit for consumer expenditure. Thus a given household can contain more than one consumer unit. This issue, and others relating to the VCB, are discussed in Officer (2007b, 2008b).

Let VCB denote the Officer series and HVCB the number of work-hours required to purchase the consumer bundle. For 1900–2006, HVCB is constructed as VCB/AHC. Table 7.3 and figure 7.4 ("Required Hours" line) present the HVCB series. Unlike the real wage, standard of living is inversely (rather than directly) related to HVCB. The fewer the number of hours to purchase the consumer bundle, the higher the workers' standard of living. There is a downward trend in HVCB until 1982, when the global minimum (1618 hours) occurs, then HVCB increases to 1865 in 1984 and remains within the 1850–2050 range thereafter.

To understand the order of magnitude of the HVCB variable, consider that a 10-hour day (achieved by 1900), 6-day week, and even 52-weeks' work together yield only 3,120 annual work-hours—exceeded by "required work-hours" until 1931. In other words, according to the HVCB measure, the standard of living of the manufacturing production worker was so low in the first three decades of the twentieth century that the fullest-time typical worker could not, by his or her own labor, purchase the consumer bundle! It is also interesting that, while AHCR increases by a multiple of 7.2 over 1900 to 2006, HVCB falls by only a factor of 0.37. For comparison with the AHCR behavior, the inverse of the 0.37 figure is 2.69. Given the criterion of purchasing power over the consumer bundle, AHCR exaggerates the improvement in standard of living by a multiple of more than two-and-a-half.

Table 7.3 Work-hours required to purchase consumer bundle: 1900–2006

Years	Number of Work-Hours									
1900–1910	5,338	5,559	5,524	5,442	5,618	5,736	5,756	5,619	5,799	6,100	6,020
1911–1920	5,833	5,857	5,659	5,625	5,385	5,248	5,261	4,657	4,358	3,651	
1921–1930	3,286	3,779	3,776	3,551	3,821	3,899	3,796	3,854	3,968	3,466	
1931–1940	3,058	2,814	2,658	2,457	2,555	2,505	2,367	2,272	2,392	2,427	
1941–1950	2,488	2,257	2,127	2,095	2,261	2,698	2,624	2,554	2,429	2,409	
1951–1960	2,284	2,233	2,214	2,230	2,287	2,249	2,267	2,172	2,241	2,221	
1961–1970	2,167	2,138	2,129	2,183	2,205	2,230	2,192	2,182	2,173	2,155	
1971–1980	2,098	2,070	1,921	1,865	1,827	1,825	1,813	1,810	1,842	1,775	
1981–1990	1,699	1,618	1,697	1,865	1,879	1,850	1,871	1,906	1,987	1,969	
1991–2000	1,983	1,909	1,904	1,916	1,936	2,007	1,921	1,954	1,974	1,965	
2001–2006	2,042	1,935	1,895	1,881	1,940	1,986					

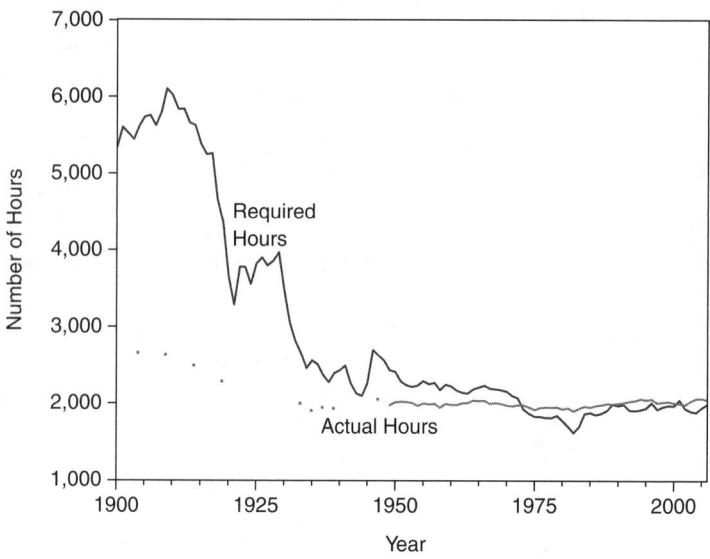

Figure 7.4 Work-hours: Actual and required-to-purchase-consumer-bundle

In the above paragraph, a *hypothetical maximum full-time* work-year provides comparison with the number of work-hours required to purchase the consumer bundle. An alternative comparison measure is the *actual* number of annual work-hours (HACT) per manufacturing production worker. Reliable figures for this variable can be constructed only for certain Census years in the twentieth century: scattered years until 1949 and then continuously. For 1904, 1909, 1914, and 1919, HACT is the product of ADO (average number of days of operation of manufacturing establishments, table 5.8) and ADH (average daily hours, using Rees figures—chapter 3, COMPOSITE SERIES). For 1933, 1935, 1937, 1939, the source of HACT is Census Man-Hour Statistics (see chapter 2, *Special Reports* under EARNINGS AND WAGES). For 1933, HACT is the sum of "average hours per wage-earner" in the twelve months; for the other years, HACT is 12 times "average hours per month." For 1947 and 1949–2006, HACT is the ratio of the total hours of production workers (source: Annual Survey of Manufactures—see chapter 5, 1920–2006) to the average number of production workers (same source).

HACT ("Actual Hours") is plotted along with HVCB in figure 7.4. While there is a downward trend in HACT, the trend ends at around 1935—because of missing observations and the limitations of

Table 7.4 Ratio of actual to consumer-bundle-required work-hours: 1904–2006

Years	Work-Hours Actual/Required Ratio									
1904–1950[a]	0.47	0.43	0.44	0.52	0.75	0.74	0.82	0.80	0.78	0.81	0.84
1951–1960	0.88	0.90	0.91	0.88	0.87	0.88	0.88	0.89	0.89	0.89	
1961–1970	0.91	0.94	0.94	0.93	0.92	0.91	0.91	0.92	0.92	0.91	
1971–1980	0.94	0.95	1.03	1.05	1.05	1.06	1.08	1.07	1.06	1.08	
1981–1990	1.14	1.17	1.14	1.05	1.04	1.06	1.06	1.05	1.00	1.02	
1991–2000	1.01	1.06	1.07	1.07	1.06	1.02	1.04	1.03	1.02	1.02	
2001–2006	0.98	1.02	1.07	1.09	1.06	1.03					

Note:
[a] Scattered years, as follows: 1904, 1909, 1914, 1919, 1933, 1935, 1937, 1939, 1947, 1949, 1950.

the Man-Hour Statistics themselves, there is an element of uncertainty here—which is much earlier than the corresponding date (1982) for HVCB.

Another innovative standard-of-living measure is the HACT/HVCB ratio: the proportion of the consumer bundle that the typical manufacturing production-worker can purchase from his or her annual earnings. This standard-of-living measure incorporates not only wage but also employment, and is shown in table 7.4. The actual/required ratio does not exceed fifty percent until 1919, though this milestone could have been reached during the war years (for which data are missing). Not until 1937 is the ratio ever above 80 percent (with the same caveat of missing observations), and the 90-percent level is reached in 1952–1953 temporarily and from 1961 continuously. Only from 1973 onward (with a slight dip in 2001) does the ratio exceed unity. Concretely, only from 1973 does the typical manufacturing production worker have sufficient annual earnings from his or her labor to purchase the entire consumer bundle. Further, in only three years (1981–1983) are annual earnings more than ten percent the cost of the consumer bundle.

It cannot be an exaggeration to state that historically the manufacturing production worker has not been a leading group among consumers in achieving enhancement of standard of living.

Standard of Living: Comparison with Other Studies

Almost every scholar who develops nominal-wage series does so with the ultimate objective of generating corresponding real-wage series or other real-wage information. Therefore application of the real average hourly compensation (AHCR) series of this study to examination of previous historical-studies' conclusions regarding the real wage is instructive. Arbitrarily, a selection is made only from historical studies published after 1965.

Adams (1968, p. 415—see chapter 2, *Antebellum Records of Firms*) examines changes in real wages in Philadelphia in 1790–1830 to state: "Two periods of rapid increase [in real wage rates] stand out—the 1790's and the period 1815–1830. The real wage increases of the 1790's were largely dissipated by 1815, but from that point on growth was the rule." Adams exhibits the average annual change in real wages of laborers (here representing unskilled occupations) and separately for five skilled occupations over 1790–1815 and 1815–1830. Taking an unweighted average of the results for the skilled occupations and combining the skilled and unskilled figures using the ten-year (1851–1860) Coelho and Shepherd (CS) Northeast weights (see chapter 5, *Interpolator and Extrapolator Series*), the average annual change in the real wage is 0.39 percent for 1790–1815 and 4.05 percent for 1815–1830.

Here the average annual percentage change in any variable Z is computed as $100 \cdot \log(Z_{t+n}/Z_t)/n$, where log represents the natural logarithm, t is the initial year, and t + n the final year. The average annual percentage change in AHCR is 0.31 percent for 1800–1815 (of necessity, replacing 1790–1815) and 4.15 percent for 1815–1830—amazingly close to the Adams figures, considering that the Adams Philadelphia data are not utilized in the present study. In all computations in this section (and, in fact, throughout the book) unrounded figures are used, resulting in superior precision to that provided by rounded figures shown in a table or stated in the text.

Putting to national use his 1821–1860 wage series based on records of civilian Army employees, Margo (2000b, p. 109—see chapter 2, *Records of Civilian Employees of U.S. Army*, and chapter 5, *Interpolator and Extrapolator Series*) estimates the annual growth rate of the U.S. real wage as the coefficient of a time trend, that is, the least-squares estimate of β in the equation $\log W_{RE} = \alpha + \beta \cdot T + \varepsilon$, where W_{RE} is the real wage, T a linear time trend, and ε an error term. Consider Margo's "variable-weights" results (which allow occupation-specific labor-force shares to vary over time in the computation of the real wage—consistent with a current-weight compensation series). Weight Margo's common-laborer and artisan growth rates according to 10-year (1851–1860) Coelho-Shepherd national weights (0.3564, 0.6436)—computed from data in CS, 1976, pp. 226, 228). Then the estimated growth rate is 0.84 percent per year. Applying the same technique and time period to AHCR, the average annual growth rate of the real wage is much greater, at 1.80 percent.

This divergence in results has several possible interpretations. It is possible that the Margo data underestimate wage growth in the economy at large; it is also possible that the AHCR series overestimates this growth. Perhaps both series are reliable; but, with the Margo series confined to males, the explosive growth in the female wage during this period (see table 5.12) is incorporated only in AHCR.

Considering the CS real-wage series (see chapter 5, *Interpolator and Extrapolator Series*, regarding the CS nominal wage), Margo (2000b, p. 9) derives an implication for real-wage behavior during the 1850s decade: "the unweighted [Coelho-Shepherd] series suggest that real wages fell during the first half of the 1850s.... Real wages then increased but were no higher in 1860 than in 1851 in any region. Thus, the Weeks Report data suggest that the 1850s was a decade of little or no overall real wage growth."

Although Margo is interpreting certain CS regional series, take here the CS (1976, p. 212) national real-wage series. This series combines all observations, unweighted across occupations and regions. For 1851–1855 the average annual growth rate is –2.29 percent; for 1856–1860 it is 1.45 percent. Corresponding figures for AHCR are –1.89 percent and 3.30 percent. Thus the AHCR series is not as pessimistic about the 1850s. In fact, while the CS national real-wage series is 2 percent lower in 1860 than in 1851, AHCR is 9 percent higher.

The CS series have an honorable but limited role in developing the AHC (and therefore AHCR) series. The methodological and data differences between AHCR and the CS series are so numerous and substantive that the differences in results are not surprising.

Margo (2006l, p. 2.44) computes a real-wage index for unskilled labor for 1774–1974. He exhibits the series not as a table but only as a graph. Both the numerator (nominal wage) and denominator (CPI) of the Margo series are series constructed by David and Solar (1977, pp. 16–17, 59–60) and reprinted in Margo (2006k) and Lindert and Sutch (2006), respectively. It is interesting that David and Solar themselves do not construct a real-wage series.

Using the time-trend regression technique, Margo estimates the average annual growth rate of the real wage for 1774–1974 (1.5 percent per year), 1774–1900 (1.2 percent per year), and 1900–1974 (2.5 percent per year). Using the same technique, but (of necessity) for 1800–1974, 1800–1900, and 1900–1974, corresponding average annual growth rates for AHCR are 2.0, 1.6, and 2.8 percent per year. Margo's (2006l, p. 2.44) statement that "two full centuries...over this very long period, real wages have increased substantially" is confirmed—even more so—via the AHCR series. Also substantiated is his observation that "the growth rate of real wages accelerated; growth was slower during the nineteenth century than in the twentieth."

The higher growth rates for AHCR are not surprising, because the David-Solar wage series pertains only to unskilled labor, whereas AHCR incorporates both skilled and unskilled workers. There are other differences between the David-Solar wage series and AHCR, but the directions of their effects are uncertain. Prior to 1890, the David-Solar data are based on unadjusted daily rather than daily-adjusted-to-hourly wage quotations; their series is occupational rather than industry based and so not specific to manufacturing; and, until 1890, their data sources are entirely different from those of AHCR. Inconsequential for the real wage but detracting from direct use is the fact that the David-Solar (nominal) wage series is an index number rather than dollar-denominated. The David-Solar wage series is discussed in David and Solar (1977, pp. 57–68) and Margo (2006k, p. 2.257). There are also conceptual and data differences between the AHCR CPI-component and the David-Solar CPI, discussed in Officer (2007a, 2008a).

The real-wage growth results of Goldin (2000, p. 565), for 1900–1929 and 1948–1973, are not considered here, because her time dimension of earnings is annual rather than daily or hourly. A comparison with AHCR growth would not be legitimate.

Margo (2006l, p. 2.44) draws the following implication from his graph of the David-Solar real-wage series: "it is apparent that year-to-year (or longer-term) variability in growth rates of real wages—volatility—was very considerable in the nineteenth century but was dampened in the twentieth century." It is not at all clear that this phenomenon is repeated in the AHCR series (figure 7.3). In particular, the first half of the twentieth century appears to exhibit cycles not present in the David-Solar series.

To examine relative volatility of the real wage in the two centuries, a technique superior to visual inspection of a graph is to use the Hodrick-Prescott filter to decompose AHCR into trend and cycle. Although Hodrick-Prescott is applied in the same way as in chapter 5, *Days of Operation*, there are two differences. First, the time period here is 1800–2006. Second, the cyclical component (CAHCR) is defined in the conventional way as AHCR –TAHCR, where TAHCR is the trend component. CAHCR is graphed in figure 7.5.

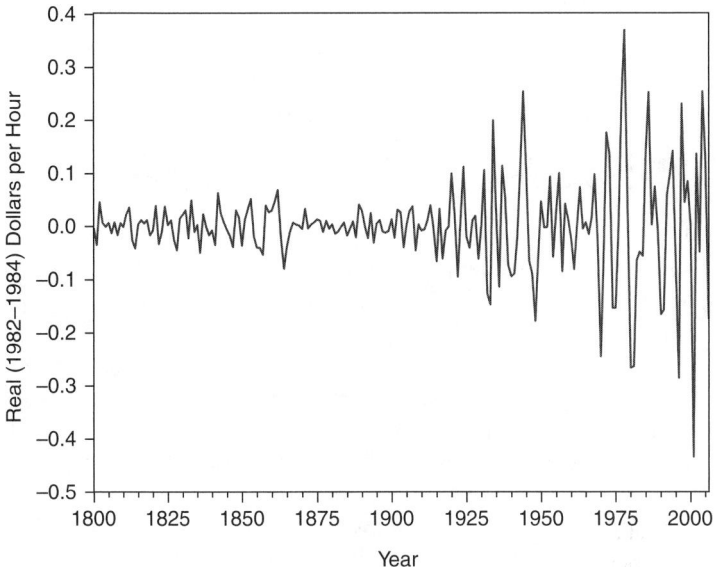

Figure 7.5 Real average hourly compensation: Cycle component

Figure 7.5 shows unambiguously that the cyclical volatility of the real wage AHCR is greater in the twentieth than the nineteenth century—the opposite of Margo's conclusion. Of course, the divergent results are due both to the differing techniques and the different real-wage series.

The relative standard deviations of CAHCR confirm the pattern in figure 7.5. For 1800–1899, the standard deviation is 0.027; for 1900–1999, it is 0.111—higher by a factor of 4.1. (The coefficient of variation [ratio of standard-deviation to mean] is not meaningful, because—inherent in the Hodrick-Prescott technique—the mean of CAHCR is zero for the entire time period [1800–2006], and therefore the mean is close to zero for the subperiods.)

In contrast, another of Margo's (2006l, p. 2.44) statements is confirmed using the AHCR series: "the so-called productivity slowdown...began about 1973. A consequence of the slowdown in productivity growth was a marked slowdown in the rate of growth of real wages." Similarly, Goldin (2000, p. 549) notes "labor productivity and real wages lagging in the United States since the mid-1970s." As evidence, Margo examines (separately) the median annual real earnings of male and female full-time workers in the entire economy for 1973–1997. Here, applying the time-trend regression technique to AHCR for 1973–1997, the estimated average annual rate of growth of the real wage is –0.12 percent. Retardation of real-wage growth during this time period applies, on average, also to manufacturing production workers (males and females together).

Concluding Comments

In summary, and notwithstanding the productivity-slowdown effect on the standard of living, two interesting results follow from this chapter's historical analysis of the standard of living of the U.S. production worker in manufacturing:

1. Applying the new series of average hourly compensation to the conventional definition of the standard of living—the real wage—the workers' standard of living exhibits greater increases than previous authors have calculated.

2. Applying the new series of average hourly compensation to original and unconventional measures of the standard of living, the increase in workers' standard of living is less impressive—much less impressive—than indicated by the real wage.

APPENDIX

Feeder Series

OCCUPATION (1880–1890): CENSUS

The number of persons engaged in each *Bulletin 18* occupation in (1879, 1889) is obtained from Census Office (1883, pp. 760–775; 1897, pp. 306–341), and in 1899 from United States Census Office (1902b, pp. 505–549). The Census occupational categories corresponding to *Bulletin 18* occupations (see chapter 5, *Interpolator and Extrapolator Series*) are shown in table A.1.

For each occupation, for (1879, 1889, 1899): the Northeast figure is the sum of the figures for Connecticut, Delaware, District of Columbia, Maine, Maryland, Massachusetts, New Hampshire, New Jersey, New York, Pennsylvania, Rhode Island, and Vermont; the rest-of-U.S. figure is the U.S. figure *minus* the Northeast figure. Intercensal values for 1880–1888 are obtained via linear interpolation between 1879 and 1889; for 1890–1898, between 1889 and 1899. Then the following series of "number of persons engaged" are obtained for 1880–1890 by simple addition: skilled occupations in the Northeast, unskilled occupations in the Northeast, skilled occupations in rest-of-U.S., unskilled occupations in rest-of-U.S. The four series serve as divisors, so that, for each occupation in each region, the "persons engaged" series for 1880–1890 is expressed as a proportion of the total number engaged in the pertinent skill group and pertinent region. These series serve as weights to construct the wage-series WSNE, WUNE, WSR, WUR (chapter 5, *Interpolator and Extrapolator Series*). Adding "number of persons engaged" in skilled and unskilled occupations for the Northeast, and similarly for rest-of-United States, and expressing the results as a proportion of the grand total, one has respective weights for *Bulletin 18* Northeast and rest-of-U.S. wages (WNEB and WRB) in the construction of the U.S. wage (WUSB)—also in chapter 5, *Interpolator and Extrapolator Series*.

Table A.1 Correspondence of *Bulletin 18* and Census occupations

Bulletin 18	Census
Skilled Occupations	
blacksmiths	blacksmiths (1899: excluding apprentices and helpers)
boiler makers	steam boiler markers (1899: excluding helpers)
cabinet makers	cabinet makers
compositors	1879: printers, lithographers, and stereotypers; 1889: printers, lithographers, and pressmen; compositors; electrotypers and stereotypers; 1899: printers, lithographers, and pressmen
iron and steel	1879: iron and steel works and shops operatives; 1889: iron and steel workers; 1899: iron and steel workers [including molders]
machinists	machinists (1899: excluding apprentices and helpers)
stonecutters	marble and stone cutters
Unskilled Occupations	
laborers, other	laborers (not specified)
teamsters	draymen, hackmen, teamsters, etc.

EMPLOYMENT

Industry

Census (1840–1890): Series of the number of wage-earners employed for eleven individual manufacturing industries over 1840–1890 are required in order to construct the Long-Aldrich and Falkner wage series (chapter 5, *Interpolator and Extrapolator Series*) and the revised Long-Aldrich average-daily-hours series [HOURS (1800–1890, later in the appendix]. The technique is to obtain employment for Census benchmark years (1840, 1849, 1859, 1869, 1879, 1889, 1899) and then linearly interpolate for the intervening years. Data sources are U.S. Census Office (1902a), the primary source for 1879, 1889, 1899; Census Office (1895), a secondary source for 1879, 1889; Walker (1872), the primary source for 1849, 1859, 1869; and Department of State (1841), the main source for 1840. Details regarding benchmark figures for the individual industries follow. Implied figures are wage-earner employment.

Agricultural implements: 1840: unnecessary, 1859: sum of 11 subindustries.

Ale, beer, porter: termed "liquors, malt." 1840: "Distilled and fermented liquors" apportioned to malt liquors via 1849 ratio of "breweries" to " 'breweries' *plus* 'distilleries' *plus* 'distilleries-rectifying.' " 1849: "breweries."

Books and newspapers: "bookbinding and blank book making" *plus* "printing and publishing." 1840: "printing and binding." 1869: "printing and publishing," sum of four components.

Carriages and wagons: 1849: not a covered industry. 1859: "carriages" *plus* "wagons and carts."

Cotton manufactures: 1859: sum of cotton "batting and wadding," "braid, thread, lines, twine, and yarn," "coverlets," "flannel carding," "goods," "mosquito-netting," and "table-cloths." 1869: sum of three sub-industries.

Illuminating gas: includes heating. 1849: assumed half of 1859. 1879: not a covered industry.

Leather: 1840: unnecessary. 1849: "tanners and curriers." 1859: sum of "leather," "leather-morocco," "leather—patent and enameled leather," and "leather—skin dressing." 1869: sum of five sub-industries.

Metals and metallic goods: represented by "iron and steel." 1840: "iron" apportioned to "iron exclusive of mining operations" via 1849 ratio of "iron" to "'iron' *plus* 'iron mining.'" 1849: sum of five iron industries, excluding "mining"; and two steel industries. 1859: sum of four iron industries, excluding "steamships"; and steel industry. 1869: sum of iron "pigs," "castings (not specified)," "blooms," and "forged and rolled"; and four steel industries.

Paper: 1840: unnecessary. 1869: sum of four sub-industries. 1899: "Paper and wood pulp" apportioned to "paper" via 1889 ratio of "paper" to "'paper' *plus* 'pulpwood.'"

White lead: White lead is an ingredient in paints, but it was also used in pottery. 1840: estimated as the product of $\{(\text{``white lead''})/(\text{``paints''} \textit{ plus } \text{``earthenware''} \textit{ plus } \text{``potteries''})\}_{1849}$ and ("earthenware, etc." *plus* "paints")$_{1840}$, where, for 1840, the "paints" components of "drugs and medicines, paints and dyes" is estimated via the 1849 ratio of "paints" to ("'paints' *plus* 'medicines, drugs, and dyestuffs'"). 1869–1899: estimated as the product of ("white lead"/"paints")$_{1859}$ to "paints," where, for 1859, "paints" are "paints" *plus* "zinc paint," and, for 1869, "paints" are "(not specified)" *plus* "lead and zinc."

Woolen manufactures: termed "woolen goods." 1849: "woolens, carding, and pulling." 1859: woolen "goods" *plus* "yarn."

Each of the 11 industry series is linearly interpolated to obtain intercensal values. In general, figures for (1841–1848, 1850–1858, 1860–1868, 1870–1878, 1880–1888, 1890) are obtained via interpolation between (1840 and 1849, 1849 and 1859, 1859 and 1869, 1869 and 1879, 1879 and 1889, 1889 and 1899). For "carriages and wagons," 1841–1858 is obtained via interpolation between 1840 and 1859; for illuminating gas, 1870–1888 via interpolation between 1869 and 1889.

Weights are then industry shares of total industry-group employment. For the Long-Aldrich wage series (RLADW, 1860–1890—chapter 5, *Interpolator and Extrapolator Series*) and the 1859–1890 component of the revised Long-Aldrich average hours per day (RLADH) series [HOURS (1800–1890)], the weight for a given industry in a given year is the proportion of the total 11-industry employment accounted for by that industry. For the Falkner index (FADW, 1859–1861—chapter 5, *Interpolator and Extrapolator Series*), the weights are the individual-industry shares of seven-industry total employment (agricultural implements, books and newspapers, carriages and wagons, leather, metals and metallic goods, white lead, woolen manufactures). For the 1857–1859 component of RLADH, the same technique is used for ten industries (with paper excluded); for 1850–1857, similarly eight industries (agricultural implements and leather also excluded); for 1840–1850, seven industries (woolen goods also excluded).

<u>Lebergott (1802–1830):</u> Lebergott (1964, p. 510) generates the number of wage-earners in the cotton-textile and the iron-and-steel industry for 1800, 1810, 1820, and 1830. The figures are reprinted in Lebergott (1966, p. 188; 1984, p. 66) and Carter (2006, p. 2110). Linear interpolation is used to create a series for 1802–1830: 1802–1809 via interpolation between 1800 and 1810, 1811–1819 between 1810 and 1820, 1821–1829 between 1820 and 1830. Dividing each series by the sum of the two series yields the employment share for each industry in the two industries combined, over 1802–1830. The textiles and iron-and-steel shares are the weights for the Adams and Zabler series in the 1802–1830 segment of the Northeast male wage series (WAZMNE—chapter 5, *Interpolator and Extrapolator Series*).

Region: Census (1820–1859)

To construct the Margo-based U.S. wage series (WUS, 1820–1859—chapter 5, *Interpolator and Extrapolator Series*), and the rest-of-U.S. average annual earnings (AAE_R, 1849) and U.S./Northeast wage ratio (RWUN; 1820, 1831, 1849)—both in chapter 5, *Average Daily Earnings*, the weights for the Northeast (E_{NE}) and rest-of-U.S. (E_R) are proportions of total U.S. manufacturing employment. Benchmark figures for 1820, 1840, 1849, and 1859 are shown in table A.2. The Northeast figure for 1859 is derived in table A.3, using the same

Table A.2 Wage-earners, manufacturing, by region, 1820–1859

Year	Number of Wage-Earners		
	United States	Northeast[a]	Rest of United States[b]
1820	349,247	218,116	131,131
1840	791,545	493,338	298,207
1849	848,668	641,184	207,484
1859	1,153,009	813,652	339,357

Notes:
[a] Sum of wage-earners in New England and Middle Atlantic states.
[b] Computed as residual.

Source: 1820 and 1840, United States and Northeast—DeBow (1854, p. 129). 1849 and 1859, United States—table 5.5. 1849, Northeast: table 5.9. 1859, Northeast—table A.3.

Table A.3 Computation of adjusted wage-earners, manufacturing, Northeast, 1859

Item	Number of Wage-Earners[a]			
	New England[b]	Extended Middle Atlantic[c]	Delaware, Maryland, DC	Northeast[d]
All industries[e]: Census	391,836	546,243	38,272	899,807
Deductions				
nonmanufacturing sectors				
agriculture	40	70	4	106
fisheries	25,312	1,452	378	26,386
forestry	765	30	30	765
mining	396	33,961	833	33,524
quarrying	2,458	2,882	139	5,201
construction	2,682	6,767	356	9,093
services	81	262	1	339
hand and custom trades				
blacksmithing	1,596	6,148	650	7,094
carving	43	143	10	176
dyeing and bleaching	40	1,035	1	1,074
kindling wood	5	387	12	380
locksmithing and bellhanging	0	134	4	130
photographs	116	393	25	484
rigging	127	167	0	294

Continued

Table A.3 Continued

Item	Number of Wage-Earners[a]			
	New England[b]	Extended Middle Atlantic[c]	Delaware, Maryland, DC	Northeast[d]
upholstery	400	741	72	1069
watch repairing	0	31[f]	0	31
whitesmithing	9	0	0	9
All industries: adjusted	357,766	491,640	35,754	813,652

Notes:
[a] Sum of "male hands employed" and "female hands employed" in source.
[b] Connecticut, Maine, Massachusetts, New Hampshire, Rhode Island, Vermont.
[c] Middle Atlantic (New Jersey, New York, Pennsylvania) and Delaware, Maryland, District of Columbia [DC].
[d] New England and Middle Atlantic.
[e] Termed "total manufactures" in source.
[f] Computed as product of 1869 U.S. ratio ("watch and clock repairing")/("watch and clock repairing" *plus* "watch materials" *plus* "watches") and 1859 extended-Middle-Atlantic sum of "watch crystals," "watch dials," "watch dials and materials," and "watches and watch repairing."

Source: New England; Extended Middle Atlantic; Delaware, Maryland, DC—all data from Secretary of the Interior (1865, pp. 55–56, 228–230, 662, 676–701). Northeast: computed as New England figure *plus* Extended Middle Atlantic figure *minus* Delaware, Maryland, DC figure.

methodology as for the computation of the number of wage-earners in the United States in table 5.5. Note that, following Margo (2000b, Table 3A.1), the strict definition of Middle Atlantic (New Jersey, New York, Pennsylvania) is adopted.

From table A.2, Northeast and rest-of-U.S. employment are each expressed as the proportion of total U.S. employment for 1820, 1840, 1849, 1859. Proportions are linearly interpolated for the intervening years: 1821–1839 via interpolation between 1820 and 1840, 1841–1848 between 1840 and 1849, 1850–1858 between 1849 and 1859. The resulting series are the required weights.

Gender (1800–1859)

Table A.4 provides estimated figures for benchmark years for the distribution of age-sex groups of workers in manufacturing in the Northeast. Only necessary cells have entries. Divide all figures by 100, so group shares are expressed as proportions of all workers. Let (E_F, E_B, E_{AM}, E_M) denote (females, boys, adult males, males) as proportions of all workers. To construct series for 1808–1859, values of

E_F and E_B are interpolated linearly between adjacent benchmark years. Then, for the interpolated years, E_{AM} is computed as $1-E_F-E_B$ and E_M calculated as $1-E_F$ or $E_{AM} + E_B$. The resulting E_F, E_B, E_{AM} figures for 1820, 1831, and 1849 are used to compute $ADE(SV)_{NE}$ (average daily earnings, all workers, Northeast, Sokoloff-Villaflor data) in equation (3) (chapter 5, *Average Daily Earnings*). The E_F and E_M series are inputs to compute the all-worker/male wage (W/W_M) for 1808–1859, in chapter 5, *Interpolator and Extrapolator Series*. For 1808–1859, the adult and boy proportions of males, denoted as $E_{AM,M}$ and $E_{B,M}$, are calculated as E_{AM}/E_M and E_B/E_M. These series are the weights for construction of the male/adult-male relative wage (W_M/W_{AM})—again in chapter 5, *Interpolator and Extrapolator Series*.

Hours (1800–1890)

The series of average daily hours (ADH) needs to be constructed for 1800–1890. ADH has three uses: computation of AHE for the benchmark years 1820, 1849, 1859, 1869, 1879, 1889 (chapter 5, *Average Daily Hours*); conversion of the Long-Aldrich daily wage series to an hourly series (ALAHW—chapter 5, *Interpolator and Extrapolator Series*); construction of the Adams-Zabler hourly wage series (WAZHMNE—again chapter 5, *Interpolator and Extrapolator Series*).

The only pre-1890 daily hours data of any reasonable length and covering multiple manufacturing industries are in the Weeks Report and the Aldrich Report. Advantages and limitations of their data are discussed in chapter 2: *Special Reports* and *Congress and Treasury*, both under EARNINGS AND WAGES. On balance, the Aldrich data are preferred, because they involve (1) annual rather than quinquennial figures, (2) average hours rather than distribution by hours intervals, and (3) scope for individual-industry weighting.

Falkner, in Aldrich Report (1893, pp. 178–179), assembles ADH series for 1840–1891 for the same industries as for his daily-wage series (see chapter 3, AVERAGE DAILY WAGE RATE). For the present study, one takes only the series for the 11 industries underlying the revised Long-Aldrich wage series (RLADW—chapter 5, *Interpolator and Extrapolator Series*) and constructs a weighted-average ADH. This is the procedure followed by Long (1960, p. 37), except that three improvements are made here: (1) industries not clearly manufacturing are excluded, whence 11 industries instead of 13; (2) time span is 1840–1890 rather than 1860–1890;

Table A.4 Estimated age-sex distribution of workers in manufacturing, Northeast[a], 1800–1859

Year	Percent of All Workers						
	Females		Children			Males[b]	
	All	Adult	All	Boys		All	Adult
1807[c]	0	0	0	0		100.0	100.0
1812[d]	6.3	—	—	3.7		93.7	90.0
1820	20.3[e]	8.9	23.1	11.7[f]		79.7	68.0
1831[g]	32.7	25.3[h]	15.0[i]	7.6		67.3	59.7
1840[j]	34.1[k]	—	—	6.9[l]		—	59.0
1849[m]	28.8	22.8[n]	12.2[o]	6.2[p]		71.2	65.0
1859[q]	28.0	22.7[r]	10.7[s]	5.4[t]		72.0	66.6

Notes:

[a] New England and Middle Atlantic states.

[b] Computed as residual or sum, by present author.

[c] Applies also to 1800–1806. Zero figure for females and boys for 1807 based on the following statements of Goldin and Sokoloff: "the burst of industrial expansion [was] ushered in by the Embargo of 1807 and the War of 1812, and it was probably during this period that the proportion of manufacturing workers composed of females and children began to increase substantially" (Goldin and Sokoloff, 1982, p. 750); "from about 1810 … the percentage of young women (between the ages of 10 and 29) engaged in factory work increased from near zero…" (Goldin and Sokoloff, 1984, p. 475). Note that the Embargo Act was passed on December 22, 1807.

[d] Combined share of females and boys based on statement: "The proportion of the northeastern manufacturing labor force composed of females and young males seems likely to have grown from about 10 percent early in the nineteenth century…" (Goldin and Sokoloff, 1982, p. 743). Division between females and boys made in proportion to 1820 figures. See also note c.

[e] Computed as sum by present author: $8.9 + (23.1 - 11.7) = 20.3$.

[f] Estimated by present author as product of (i) 1831 ratio of boys to all children and (ii) all children in 1820: $(7.6/15.0) \cdot 23.1 = 11.7$.

[g] Stated as 1832 in source.

h Girls constituted 20–25 percent of the total female workforce in manufacturing in 1831 (Goldin and Sokoloff, 1984, p. 480). Taking the midpoint of that range, 22.5 percent of 32.7 yields 7.4 percent of all workers, leaving adult females accounting for 25.3 percent of all workers.

i Sum of 7.6 (boys) and 7.4 (girls, computed in note h).

j Estimated peak year for combined female and boy share of manufacturing workforce: "women and children... their fraction of the manufacturing labor force in the Northeast... achieving an historical peak in the vicinity of 40 percent sometime between those years [1820 and 1850]... cresting near 40 percent.... The likelihood [is] that the peak occurred during the late 1830s or early 1840s..." Goldin and Sokoloff (1982, pp. 746–747). It follows that 1840 is the logical estimate of the peak year. With females and boys constituting 40.3 percent of manufacturing workers in an earlier year (1831), a peak value of 41.0 percent (in 1840) is reasonable.

k Computed as difference between 41.0 (see note j) and 6.9 (boy share).

l Computed via linear interpolation between 1831 and 1849

m Stated as 1850 in source. Figures include "clothiers and tailors," consistent with present-study benchmark figures for 1849.

n Computed as residual: $28.8 - (12.2 - 6.2) = 22.8$.

o Estimated by present author as product of (i) 1831 ratio of all children to boys and (ii) boys in 1849: $(15.0/7.6) \cdot 6.2 = 12.2$.

p Goldin and Sokoloff (1984, p. 476, n. 16) state a figure of 4.6 (misprinted as 3.6) percent for 1870, and, for 1850, they take the midpoint of 4.6 and 7.6, the 1831 (for them, 1832) figure. Strict linear interpolation yields a figure of 6.2 for 1849 relative to 1831 and 1869 (the applicable Census calendar year).

q Stated as 1860 in source.

r Computed as residual: $28.0 - (10.7 - 5.4) = 22.7$.

s Estimated by present author as product of (i) 1831 ratio of all children to boys and (ii) boys in 1859: $(15.0/7.6) \cdot 5.4 = 10.7$.

t Figure obtained via linear interpolation according to the procedure in note p.

Source: Goldin and Sokoloff (1982, pp 743, 748–749), except where otherwise noted.

(3) weighting pattern is transparent, with judgments explicit (see *Industry*).

For "ale, beer, and porter," the missing years (1840–1842, 1845–1853) are given the figure of 12.0, which is the value for all other years. For "books and newspapers," the missing years (1840–1841) are given the figure of 10.0, which applies to all other years. For "white lead," the missing year (1841) is given the figure 9.3, which is the value for 1840 and 1842–1862. Then there are continuous data for all 11 industries for 1859–1890, for ten industries (paper excluded) for 1857–1859, for eight industries (agricultural implements and leather also excluded) for 1850–1857, and for seven industries (woolen goods also excluded) for 1840–1850. An employment-weighted average series is constructed for each of the four time periods using the weights derived in *Industry*.

A revised Long-Aldrich average daily hours (RLADH) series is derived for 1831–1890 as follows. For 1859–1890, RLADH is the 11-industry series. For 1857–1858, RLADH is the 10-industry series multiplied by the 1859 ratio of RLADH to the 10-industry series; for 1850–1856, the eight-industry series multiplied by the 1857 ratio of RLADH to the eight-industry series; for 1840–1849, the seven-industry series multiplied by the 1850 ratio of RLADH to the seven-industry series. RLADH is extended to 1831 via Weeks data. Using the Sundstrom (2006b) technique of averaging the hour-intervals lower-bounds weighted by the number of statements, a constant figure is obtained for 1830, 1835 and 1840 (see final column of table A.5). This result justifies extrapolating the 1840 value of RLADH to 1831–1839.

Benchmark values of the final ADH series exist for 1831, 1879–1880, and 1890 (see chapter 5, *Average Daily Hours*). ADH is estimated for 1832–1878 and 1881–1889 using the adopted interpolating method (see chapter 5, 1920–2006) with ADH the desired series and RLADH the interpolator series. Beginning and ending years of the interpolation are 1831 and 1879 for 1832–1878, 1880 and 1890 for 1881–1889. Thus ADH has been constructed for 1831–1890.

The ADH figure for 1831 is extrapolated back to 1800. The conventional wisdom is different and is stated succinctly by Wright (1885, p. 10): "The hours of labor in nearly all industries were measured by the sun, from sunrise to sunset constituting the working day. Not...until 1840 were shorter hours adopted to any extent." "Sunrise to sunset" connotes an average twelve-hour day over the year. Consistent with this view, Lebergott (1964, p. 48) refers to the

Table A.5 Average hours per day, production workers[a] in manufacturing: Comparison of new series with existing series, 1830–1890

Year	New Series	Existing Series				
		Long		Whaples[b,d,e]	Margo[e,f]	Sundstrom[g]
		Weeks[b]	Aldrich[c]			
1830	11.33	—	—	11.5	11.5	10.9
1831	11.33[h]	—	—	—	—	—
1832	11.34	—	—	—	—	—
1833	11.36	—	—	—	—	—
1834	11.37	—	—	—	—	—
1835	11.38	—	—	—	—	10.9
1836	11.40	—	—	—	—	—
1837	11.41	—	—	—	—	—
1838	11.42	—	—	—	—	—
1839	11.44	—	—	—	—	—
1840	11.45	—	—	11.2	11.3	10.9
1841	11.45	—	—	—	—	—
1842	11.45	—	—	—	—	—
1843	11.26	—	—	—	—	—
1844	11.33	—	—	—	—	—
1845	11.41	—	—	—	—	10.7
1846	11.37	—	—	—	—	—
1847	11.35	—	—	—	—	—
1848	11.34	—	—	—	—	—
1849	11.32	—	—	—	—	—
1850	11.32	—	—	10.9	11.2	10.6
1851	11.16	—	—	—	—	—
1852	10.85	—	—	—	—	—
1853	10.81	—	—	—	—	—
1854	10.78	—	—	—	—	—
1855	10.75	—	—	—	—	10.4
1856	10.68	—	—	—	—	—
1857	10.67	—	—	—	—	—
1858	10.67	—	—	—	—	—
1859	10.64	—	—	—	—	—
1860	10.59	10.9	10.8	10.3	10.7	10.4
1861	10.39	—	10.7	—	—	—
1862	10.33	—	10.7	—	—	—
1863	10.31	—	10.7	—	—	—
1864	10.31	—	10.7	—	—	—
1865	10.23	10.9	10.6	—	—	10.4
1866	10.29	—	10.7	—	—	—
1867	10.31	—	10.7	—	—	—
1868	10.14	—	10.6	—	—	—
1869	10.15	—	10.6	—	—	—
1870	10.13	10.8	10.5	10.2	10.4	10.3

Continued

Table A.5 Continued

Year	New Series	Existing Series				
		Long		Whaples[b,d,e]	Margo[e,f]	Sundstrom[g]
		Weeks[b]	Aldrich[c]			
1871	10.14	—	10.5	—	—	—
1872	10.16	—	10.5	—	—	—
1873	10.17	—	10.5	—	—	—
1874	10.18	—	10.5	—	—	—
1875	10.01	10.8	10.4	—	—	10.3
1876	10.01	—	10.4	—	—	—
1877	10.05	—	10.4	—	—	—
1878	10.04	—	10.4	—	—	—
1879	10.05	—	10.4	—	—	—
1880	10.05	10.8	10.4	10.1	10.2	10.3
1881	10.07	—	10.4	—	—	—
1882	10.08	—	10.4	—	—	—
1883	10.10	—	10.3	—	—	—
1884	10.11	—	10.3	—	—	—
1885	10.13	—	10.3	—	—	—
1886	10.05	—	10.2	—	—	—
1887	9.95	—	10.0	—	—	—
1888	9.97	—	10.0	—	—	—
1889	10.00	—	10.0	—	—	—
1890	10.02	—	10.0	—	10.0	—

Notes:
[a] Also termed "wage-earners" or "manual workers."
[b] Based on Weeks Report.
[c] Based on Aldrich Report.
[d] Method of computation from Weeks table unstated. Also provides Aldrich Report figures at 10-year intervals.
[e] Weekly hours in source; daily hours obtained by division by 6.
[f] Based on various series in Whaples (1990, p. 33).
[g] Based on Weeks Report. Sundtrom provides figures only for 1830 and 1880; other years computed by present author using Sundstrom's technique of weighting lower bounds of work-hour intervals by number of statements.
[h] Figure extended back to 1800.

Source: Long (1960, pp. 35, 37), Whaples (1990, p. 33), Margo (2000a, p. 230), Sundstrom (2006b). For "New Series," see text.

Weeks Report for evidence of an average workday of 13 hours "in the dominant textile industries" and 11–12 hours "in the others." Whaples (2001a, p. 8) notes "the common working day of twelve hours" in textiles. The problem with the conventional wisdom is that it contradicts the McLane Report. On the basis of data for many manufacturing industries, the McLane Report indicates a shorter average workday, 11.33 hours, in 1831 (see chapter 5, *Average Daily Hours*). Even if at one time 12 hours were the norm, it is unclear how to integrate that information with the firmly based 1831 figure. Extrapolation appears to be the appropriate course. Table A.5 lists the ADH series ("New Series") along with the hours series of other private scholars.

REFERENCES

Abbott, Edith (1905). "The Wages of Unskilled Labor in the United States, 1850–1900." *Journal of Political Economy* 13 (June), pp. 321–367.
Abraham, Katharine G., James R. Spletzer, and Jay C. Stewart (1998). "Divergent Trends in Alternative Wage Series." In John Haltiwanger, Marilyn E. Manser, and Robert Topel, eds., *Labor Statistics Measurement Issues,* pp. 293–324. Chicago: University of Chicago Press.
Achinstein, Ascher (1930). Review of Paul F. Brissenden, *Earnings of Factory Workers, 1899 to 1927. Journal of the American Statistical Association* 25 (September), pp. 370–371.
Adams, Donald R., Jr. (1967). *Wage Rates in Philadelphia, 1790–1830.* Ph.D. dissertation. Philadelphia: University of Pennsylvania. Reprinted by Arno Press, 1975.
——— (1968). "Wage Rates in the Early National Period: Philadelphia, 1785–1830." *Journal of Economic History* 28 (September), pp. 404–426.
——— (1973). "Wage Rates in the Iron Industry: A Comment." *Explorations in Economic History* 11 (Fall), pp. 90–94.
——— (1982). "The Standard of Living during American Industrialization: Evidence from the Brandywine Region, 1800–1860." *Journal of Economic History* 42 (December), pp. 903–917.
[Aldrich Report] Aldrich, Nelson W. (1893). *Wholesale Prices, Wages, and Transportation.* Senate Report 1394, Part 1. 52nd Cong., 2nd Sess. Washington, DC: Government Printing Office.
Allen, Steven G. (1992). "Changes in the Cyclical Sensitivity of Wages in the United States, 1891–1987." *American Economic Review* 82 (March), pp. 122–140.
American FactFinder Help (undated-a). "General Statistics." U.S. Census Bureau. *2006 Annual Survey of Manufactures.* URL: http://factfinder.census.gov/servlet.
——— (undated-b). "Scope." U.S. Census Bureau. *2006 Annual Survey of Manufactures.* URL: http://factfinder.census.gov/servlet.
Atack, Jeremy (1987). "Economies of Scale and Efficiency Gains in the Rise of the Factory in America, 1820–1900." In Peter Kilby, ed., *Quantity and Quiddity: Essays in U.S. Economic History,* pp. 286–335. Middletown, CT: Wesleyan University Press.
Atack, Jeremy, and Fred Bateman (1992). "How Long Was the Workday in 1880?" *Journal of Economic History* 52 (March), pp. 129–160.

Atack, Jeremy, and Fred Bateman (1995). *Irregular Production and Time-Out-of-Work in American Manufacturing Industry in 1870 and 1880: Some Preliminary Estimates.* Historical Paper No. 69. Cambridge, MA: National Bureau of Economic Research.

——— (1999). "Nineteenth-Century U.S. Industrial Development through the Eyes of the Census of Manufactures." *Historical Methods* 32 (Fall), pp. 177–188.

Atack, Jeremy, Fred Bateman, and Robert A. Margo [ABM] (2002). "Part-Year Operation in Nineteenth-Century American Manufacturing: Evidence from the 1870 and 1880 Censuses." *Journal of Economic History* 62 (September), pp. 792–809.

——— (2003). "Productivity in Manufacturing and the Length of the Working Day: Evidence from the 1880 Census of Manufactures." *Explorations in Economic History* 40, pp. 170–194.

Barkume, Anthony J., and Michael K. Lettau (2000). "Replicate Estimates of the Average Hourly Earnings Series." *Monthly Labor Review* 123 (October), pp. 12–18.

Barnett, George E. (1924). Review of Willford I. King, *Employment, Hours, and Earnings in Prosperity and Depression: United States, 1920–1922. Journal of the American Statistical Association* 19 (March), pp. 108–110.

Beney, M. Ada (1936). *Wages, Hours, and Employment in the United States, 1914–1936.* New York: National Industrial Conference Board.

Bowden, Witt (1955a). "BLS Historical Estimates of Earnings, Wages, and Hours." *Monthly Labor Review* 78 (July), pp. 801–806.

——— (1955b). "Nongovernmental Historical Series on Earnings, Wages, and Hours." *Monthly Labor Review* 78 (August), pp. 918–921.

Bowley, Arthur L. (1895). "Comparison of the Rates of Increase of Wages in the United States and in Great Britain, 1860–1891." *Economic Journal* 5 (September), pp. 369–383.

Brandes, Stuart D. (1976). *American Welfare Capitalism, 1880–1940.* Chicago: University of Chicago Press.

Brissenden, Paul F. (1929). *Earnings of Factory Workers, 1899 to 1927: An Analysis of Pay-Roll Statistics.* Census Monograph 10. Washington, DC: Government Printing Office.

[*Bulletin 18*] Department of Labor (1898). *Wages in the United States and Europe, 1870 to 1898.* Bulletin of the Department of Labor, No. 18. Washington, DC: Government Printing Office.

[*Bulletin 77*] Bureau of Labor (1908). "Wages and Hours of Labor in Manufacturing Industries, 1890 to 1907." *Bulletin of the Bureau of Labor* No. 77 (July). Washington, DC: Government Printing Office.

[*Bulletin 93*] Bureau of the Census (1908). *Earnings of Wage-Earners.* Census of Manufactures, 1905. Washington, DC: Government Printing Office.

Bullock, Charles J. (1899). "Wage Statistics and the Federal Census." *Publications of the American Economic Association*, New Series, No. 2, The Federal Census. Critical Essays by Members of the American Economic Association (March), pp. 343–368.

Bureau of the Census (1907). *Manufactures, 1905*, Part 1. Washington, DC: Government Printing Office.
——— (1913). *Manufactures: 1909*, vol. 8. Thirteenth Census. Washington, DC: Government Printing Office.
——— (1917). *Abstract of the Census of Manufactures, 1914.* Washington, DC: Government Printing Office.
——— (1923). *Manufactures: 1919*, vol. 8. Fourteenth Census. Washington, DC: Government Printing Office.
——— (1924). *Census of Manufactures, 1921.* Washington, DC: Government Printing Office.
——— (1926). *Biennial Census of Manufactures, 1923.* Washington, DC: Government Printing Office.
——— (1933). *Manufactures: 1929*, vol. 1. Fifteenth Census. Washington, DC: Government Printing Office.
——— (1935). *Man-Hour Statistics for 32 Selected Industries.* Census of Manufactures: 1933.
——— (1938). *Man-Hour Statistics for 59 Selected Industries.* Census of Manufactures: 1935.
——— (1939). *Man-Hour Statistics for 105 Selected Industries.* Census of Manufactures: 1937.
——— (1942). *Man-Hour Statistics for 171 Selected Industries.* Census of Manufactures: 1939.
——— (1949). *Historical Statistics of the United States, 1789–1945.* Washington, DC: Government Printing Office.
——— (1960). *Historical Statistics of the United States, Colonial Times to 1957.* Washington, DC: Government Printing Office.
——— (1975). *Historical Statistics of the United States, Colonial Times to 1970, Bicentennial Edition*, 2 vols. Washington, DC: Government Printing Office.
——— (1981). *Subject Statistics.* 1977 Census of Manufactures, vol. 1.
——— (1986). *Statistics for Industry Groups and Industries.* 1984 Annual Survey of Manufactures.
——— (1998). *Statistics for Industry Groups and Industries.* 1996 Annual Survey of Manufactures.
Bureau of Labor Statistics [BLS] (1927). "Workmen's Compensation and Social Insurance." *Monthly Labor Review* 24 (June 1927), pp. 76–86.
——— (1960). *Composition of Payroll Hours in Manufacturing, 1958.* Bulletin 1283.
——— (1969). *Employee Compensation in the Private Nonfarm Economy, 1966.* Bulletin 1627.
——— (1971). *Employee Compensation in the Private Nonfarm Economy, 1968.* Bulletin 1722.
——— (1973). *Employee Compensation in the Private Nonfarm Economy, 1970.* Bulletin 1770.
——— (1975). *Employee Compensation in the Private Nonfarm Economy, 1972.* Bulletin 1873.

Bureau of Labor Statistics (1976a). *BLS Handbook of Methods for Surveys and Studies*. Bulletin 1910.

——— (1976b). *The Hourly Earnings Index, 1964–August 1975*. Bulletin 1897.

——— (1977). *Employee Compensation in the Private Nonfarm Economy, 1974*. Bulletin 1863.

——— (1980). *Handbook of Labor Statistics*. Bulletin 2070.

——— (1981, 1983, 1988, 1989). *Current Wage Developments*, various issues.

——— (1985). *Handbook of Labor Statistics*. Bulletin 2217.

——— (1986). *BLS Measures of Compensation*. Bulletin 2239.

——— (1988). *BLS Handbook of Methods*. Bulletin 2285.

——— (1996). *1995 Hours at Work Survey*.

——— (1997). "Productivity Measures: Business Sector and Major Subsectors." *BLS Handbook of Methods*, Chapter 10.

——— (2000a). *Employer Costs for Employee Compensation, 1986–99*. Bulletin 2526.

——— (2000b). *Employment Cost Indexes, 1975–99*. Bulletin 2532.

——— (2001). *2000 Hours at Work Survey*.

——— (2002). *Employer Costs for Employee Compensation, Historical Listing (Annual), 1986–2001*.

——— (2003a). *Industry Analytical Ratios for [the—sic] Manufacturing, All Persons*.

——— (2003b). *A Note on the Incorporation of Hours-Worked Hours-Paid Ratios from the Employment Cost Index into Hours at Work Measures*.

Bureau of Labor Statistics (2006a). *Changes to the Current Employment Statistics Survey*. URL: http://www.bls.gov/ces/cesww.htm.

——— (2006b). *Employer Costs for Employee Compensation—December 2006*.

——— (updated). "Employment, Hours, and Earnings from the Establishment Survey." In *BLS Handbook of Methods*, Chapter 2.

——— (updated). "National Compensation Measures." In *BLS Handbook of Methods*, Chapter 8.

Calomiris, Charles W., and Christopher Hanes (1994). "Consistent Output Series for the Antebellum and Postbellum Periods: Issues and Preliminary Results." *Journal of Economic History* 54 (June), pp. 409–422.

Caroll, Richard E. (2006). "Changes Affecting the Employment Cost Index: An Overview." *Monthly Labor Review* 129 (April), pp. 3–5.

Carter, Susan B. (2006). "The Labor Force, by Industry, 1800–1960 [Lebergott and Weiss]." In Carter et al. (2006), pp. 2.110–111.

Carter, Susan B., Scott Sigmund Gartner, Michael R. Haines, Alan L. Olmstead, Richard Sutch, and Gavin Wright (2006). *Historical Statistics of the United States, Earliest Times to the Present, Millennial Edition*, 5 vols. New York: Cambridge University Press.

Carter, Susan B., John A. James, and Richard Sutch (2006). "Net National Saving—Person, Corporate, and Government: 1897–1949 [Goldsmith]." In Carter et al. (2006), pp. 3.298–299.

Census Office (1860). *Instructions to U.S. Marshals*. Eighth Census. Washington, DC: Geo. W. Bowman.

——— (1883). *Statistics of the Population of the United States*. Tenth Census. Washington, DC: Government Printing Office.

——— (1895). *Report on Manufacturing Industries in the United States*, Part I, Totals for States and Industries. Eleventh Census. Washington, DC: Government Printing Office.

——— (1897). *Report on Population of the United States*, Part 2. Eleventh Census. Washington, DC: Government Printing Office.

Chandler, Alfred D., Jr. (1977). *The Visible Hand: The Managerial Revolution in American Business*. Cambridge, MA: Harvard University Press.

Coelho, Philip R. P., and James F. Shepherd [CS] (1976). "Regional Differences in Real Wages: The United States, 1851–1880." *Journal of Economic History* 13 (April), pp. 203–230.

Commissioner of Labor (1886). *First Annual Report, Industrial Depressions*. Washington: Government Printing Office. Reprinted in *Report of the Secretary of the Interior*, 49th Cong., 1st Sess., House of Representatives, Executive Document 1, pt. 5.

——— (1905). *Wages and Hours of Labor*. Nineteenth Annual Report, 1904. House of Representatives, Document No. 428, 58th Cong. 3rd Sess. Washington, DC: Government Printing Office.

Coombs, Whitney (1926). *The Wages of Unskilled Labor in Manufacturing Industries in the United States, 1890–1924*. New York: Columbia University Press.

Costo, Stephanie L. (2006). "Introducing 2002 Weights for the Employment Cost Index." *Monthly Labor Review* 129 (April), pp. 28–32.

Creamer, Daniel (1950). *Behavior of Wage Rates during Business Cycles*. Occasional Paper 34. New York: National Bureau of Economic Research.

David, Paul A., and Peter Solar (1977). "A Bicentenary Contribution to the History of the Cost of Living in America." In Paul Uselding, ed., *Research in Economic History* 2, pp. 1–80. Greenwich, CT: JAI Press.

Davis, Joseph H. (2004). "An Annual Index of U.S. Industrial Production, 1790–1815." *Quarterly Journal of Economics* 119 (November), pp. 1177–1215.

DeBow, James Dunwoody Brownson (1853). *The Seventh Census of the United States: 1850*. Appendix. Washington, DC: Robert Armstrong.

——— (1854). *Statistical View of the United States*. Seventh Census Washington, DC: Beverley Tucker.

Department of State (1841). *Compendium of the Enumeration of the Inhabitants and Statistics of the United States*. Sixth Census. Washington, DC: Thomas Allen.

Devens, Richard M., Jr. (1978). "The Average Workweek: Two Surveys Compared." *Monthly Labor Review* 101 (July), pp. 3–8.

Dewey, Davis R. (1903). *Employees and Wages*. Twelfth Census of the United States Taken in the Year 1900. Special Reports. Washington, DC: U.S. Census Office.

Douglas, Paul H. (1930). *Real Wages in the United States, 1890–1926*. Boston: Houghton.

Douglas, Paul H. (1962). Review of Albert Rees, *Real Wages in Manufacturing, 1890–1914. American Historical Review* 67 (January), pp. 446–448.

Douglas, Paul H., and Frances Lamberson (1921). "The Movement of Real Wages, 1890–1918." *American Economic Review* (September), pp. 409–426.

Douty, Harry M. (1984). "A Century of Wage Statistics: the BLS Contribution." *Monthly Labor Review* 107 (November), pp. 16–28.

Easterlin, Richard A. (1957). "Estimates of Manufacturing Activity." In Simon Kuznets and Dorothy Swaine Thomas, eds., *Population Redistribution and Economic Growth: United States, 1870–1950*, pp. 635–701. Philadelphia, PA: American Philosophical Society.

Fabricant, Solomon (1942). *Employment in Manufacturing, 1899–1939: An Analysis of Its Relation to the Volume of Production*. New York: National Bureau of Economic Research.

Famulari, Melissa, and Marilyn E. Manser (1989). "Employer-Provided Benefits: Employer Cost versus Employee Value." *Monthly Labor Review* 112 (December), pp. 24–32.

Fishback, Price V., and Shawn Everett Kantor (1996). "The Durable Experiment: State Insurance of Workers' Compensation Risk in the Early Twentieth Century." *Journal of Economic History* 56 (December), pp. 809–836.

——— (1998). "The Adoption of Workers' Compensation in the United States, 1900–1930." *Journal of Law and Economics* (October), pp. 305–341.

——— (2000). *A Prelude to the Welfare State: The Origins of Workers' Compensation*. Chicago: University of Chicago Press.

Fishback, Price V., and Melissa A. Thomasson (2006). "Social Welfare: 1929 to the Present." In Carter et al. (2006), pp. 2.700–719.

Frickey, Edwin (1942). *Economic Fluctuations in the United States: A Systematic Analysis of Long-Run Trends and Business Cycles, 1866–1914*. Cambridge, MA: Harvard University Press.

——— (1947). *Production in the United States, 1860–1914*. Cambridge, MA: Harvard University Press.

Gallman, Robert E. (1960). "Commodity Output, 1839–1899." In Conference on Research in Income and Wealth, *Trends in the American Economy in the Nineteenth Century*, pp. 13–67. National Bureau of Economic Research. Princeton: Princeton University Press.

Getz, Patricia M. (2003). "CES Program: Changes Planned for Hours and Earnings Series." *Monthly Labor Review* 126 (October), pp. 38–39.

Goldin, Claudia (1990). *Understanding the Gender Gap: An Economic History of American Women*. New York: Oxford University Press.

——— (2000). "Labor Markets in the Twentieth Century." In Stanley L. Engerman and Robert E. Gallman, eds., *The Cambridge Economic History of the United States*, vol. 3, *The Twentieth Century*, pp. 549–623. New York: Cambridge University Press.

Goldin, Claudia, and Kenneth Sokoloff (1982). "Women, Children, and Industrialization in the Early Republic: Evidence from the Manufacturing Censuses." *Journal of Economic History* 42 (December), pp. 741–774.

——— (1984). "The Relative Productivity Hypothesis of Industrialization: The American Case, 1820 to 1850." *Quarterly Journal of Economics* 99 (August), pp. 461–487.

Greis, Theresa Diss (1984). *The Decline of Annual Hours Worked in the United States since 1947.* Philadelphia: Industrial Research Unit, University of Pennsylvania.

Grosse, Scott D. (1982). "On the Alleged Antebellum Surge in Wage Differentials: A Critique of Williamson and Lindert." *Journal of Economic History* 42 (June), pp. 413–418.

Haines, Michael R. (2006). "State Populations." In Carter et al. (2006), pp. 1.180–379.

Hanes, Christopher (1992). "Comparable Indices of Wholesale Prices and Manufacturing Wage Rates in the United States, 1865–1914." In Roger L. Ransom, Richard Sutch, and Susan B. Carter, eds., *Research in Economic History* 14, pp. 269–292. Greenwich, CT: JAI Press.

——— (1993). "The Development of Nominal Wage Rigidity in the Late 19th Century." *American Economic Review* 83 (September), pp. 732–756.

——— (1996). "Changes in the Cyclical Behavior of Real Wage Rates, 1870–1990." *Journal of Economic History* 56 (December), pp. 837–861.

Hansen, Alvin H. (1925). "Factors Affecting the Trend of Real Wages." *American Economic Review* 15 (March), pp. 27–42.

Jablonski, Mary, Kent Kunze, and Phyllis Flohr Otto [JKO] (1990). "Hours at Work: A New Base for BLS Productivity Statistics." *Monthly Labor Review* (February), pp. 17–24.

Jacobs, Eva. E., ed. (2005). *Handbook of U.S. Labor Statistics*, 8th edition. Lanham, MD: Bernan.

James, John A., and Richard Sylla (2006). "Personal Saving, by Major Components of Assets and Liabilities: 1897–1949 [Goldsmith]." In Carter et al. (2006), pp. 3.300–303.

Johnston, Louis D., and Samuel H. Williamson (2008). *What Was the U.S. GDP Then?* MeasuringWorth. URL: http://www.measuringworth.org/usgdp.

Jones, Ethel B. (1961). *Hours of Work in the United States, 1900–1957.* Ph.D. dissertation. Chicago: University of Chicago.

——— (1963). "New Estimates of Hours of Work per Week and Hourly Earnings, 1900–1957." *Review of Economics and Statistics* (November), pp. 374–385.

——— (1974). *An Investigation of the Stability of Hours of Work per Week in Manufacturing, 1947–1970.* Research Monograph No. 7. Athens, GA: Division of Research, College of Business Administration, University of Georgia.

Kendrick, John W. (1961). *Productivity Trends in the United States.* National Bureau of Economic Research. Princeton: Princeton University Press.

Kennedy, Joseph C. G. (1859). *Abstract of the Statistics of Manufactures according to the Returns of the Seventh Census.* Senate Executive Document No. 39, 35th Congress, 2nd Sess.

King, Willford Isbell (1923). *Employment Hours and Earnings in Prosperity and Depression: United States, 1920–1922,* 2nd edition. New York: National Bureau of Economic Research.

Kirkland, Katie (2000). "On the Decline in Average Weekly Hours Worked." *Monthly Labor Review* 123 (July), pp. 26–31.

Kunze, Kent (1984). "A New BLS Survey Measures the Ratio of Hours Worked to Hours Paid." *Monthly Labor Review* 107 (June), pp. 3–7.

——— (1985). "Hours of Work Increase Relative to Hours Paid." *Monthly Labor Review* 108 (June), pp. 44–46.

Latimer, Murray W. (1930). Review of Paul H. Douglas, *Real Wages in the United States, 1890–1926. Journal of the American Statistical Association* 25 (December), pp. 479–485.

Layer, Robert G. (1955). *Earnings of Cotton Mill Operatives, 1825–1914.* Committee on Research in Economic History. Cambridge, MA: Harvard University Press.

Lebergott, Stanley (1961a). Review of Albert Rees, *Real Wages in Manufacturing, 1890–1914. American Economic Review* 51 (September), pp. 773–774.

——— (1961b). Review of Clarence D. Long, *Wages and Earnings in the United States, 1860–1890. Journal of Economic History* 21 (June), pp. 262–264.

——— (1964). *Manpower in Economic Growth: The American Record since 1800.* New York: McGraw-Hill.

——— (1966). "Labor Force and Employment, 1800–1960." In Conference on Research in Income and Wealth, *Output, Employment, and Productivity in the United States after 1800,* pp. 117–204. National Bureau of Economic Research. New York: Columbia University Press.

——— (1984). *The Americans: An Economic Record.* New York: W. W. Norton.

Lindert, Peter H., and Richard Sutch (2006). "Consumer Price Indexes, for All Items: 1774–2003." In Carter et al. (2006), pp. 3.158–159.

Lindert, Peter H., and Jeffrey G. Williamson (1982). "Antebellum Wage Widening Once Again." *Journal of Economic History* 42 (June), pp. 419–422.

Long, Clarence D. (1960). *Wages and Earnings in the United States, 1860–1890.* National Bureau of Economic Research. Princeton: Princeton University Press.

Magdoff, Harry, Irving H. Siegel, and Milton B. Davis (1939). *Production, Employment, and Productivity in 59 Manufacturing Industries.* Philadelphia: Works Progress Administration, National Research Project.

Margo, Robert A. (1992). "Wages and Prices during the Antebellum Period: A Survey and New Evidence." In Robert E. Gallman and John Joseph Wallis, eds., *American Economic Growth and Standards of Living before the Civil War,* pp. 173–216. Chicago: University of Chicago Press.

——— (1998). "Wages and Labor Markets before the Civil War." *American Economic Review* 88 (May), pp. 51–56.

——— (2000a). "The Labor Force in the Nineteenth Century." In Stanley L. Engerman and Robert E. Gallman, eds., *The Cambridge Economic History of the United States*, vol. 2, *The Long Nineteenth Century*, pp. 207–243. New York: Cambridge University Press.

——— (2000b). *Wages and Labor Markets in the United States, 1820–1860.* Chicago: University of Chicago Press.

——— (2006a). "Annual and Hourly Earnings in Manufacturing, by Industry: 1889–1914." In Carter et al. (2006), pp. 2.268–269.

——— (2006b). "Annual Earnings in Selected Industries and Occupations: 1890–1926." In Carter et al. (2006), pp. 2.271–272.

——— (2006c). "Annual Earnings of Male Manufacturing Workers in New England and the Middle Atlantic, by Urban-Rural Location: 1820–1860." In Carter et al. (2006), p. 2.261.

——— (2006d). "Daily and Monthly Wages for Common Labor, Artisans, and Clerks, by Region: 1821–1860." In Carter et al. (2006), pp. 2.262–263.

——— (2006e). "Daily and Monthly Wages in the Philadelphia and Brandywine Regions, 1785–1960." In Carter et al. (2006), p. 2.258.

——— (2006f). "Daily Wages for Common Labor, by Region: 1851–1880." In Carter et al. (2006), p. 2.264.

——— (2006g). "Hourly and Weekly Earnings in Selected Industries and for Lower Skilled Labor: 1890–1926." In Carter et al. (2006), p. 2.270.

——— (2006h). "Hourly and Weekly Earnings of Production Workers in Manufacturing, by Sex and Degree of Skill: 1914–1948." In Carter et al. (2006), p. 2.279.

——— (2006i). "Hourly Earnings in Manufacturing: 1923–1990." In Carter et al. (2006), p. 2.281.

——— (2006j). "Hourly Wages in Manufacturing, by Industry, 1865–1914." In Carter et al. (2006), p. 2.267.

——— (2006k). "Index of Money Wages for Unskilled Labor: 1774–1974." In Carter et al. (2006), pp. 2.256–257.

——— (2006l). "Wages and Wage Inequality." In Carter et al. (2006), pp. 2.40–46.

Margo, Robert A., and Georgia C. Villaflor (1987). "The Growth of Wages in Antebellum America: New Evidence." *Journal of Economic History* 47 (December), pp. 873–895.

[McLane Report] Secretary of the Treasury (1833). *Documents Relative to the Manufactures in the United States,* 2 vols. House of Representatives. 22nd Cong., 1st Sess. Washington, DC: Duff Green.

Millis, Harry A., and Royal E. Montgomery (1938). *Labor's Risks and Social Insurance.* New York: McGraw-Hill.

Miron, Jeffrey A., and Christina D. Romer (1990). "A New Monthly Index of Industrial Production, 1884–1910." *Journal of Economic History* 50 (June), pp. 321–337.

Mitchell, Wesley C. (1908). *Gold, Prices, and Wages under the Greenback Standard.* Berkeley: Berkeley University Press.

Moehrle, Thomas G. (2001). "The Evolution of Compensation in a Changing Economy." *Compensation and Working Conditions* (Fall), pp. 9–15.

Morisi, Teresa L. (2003). "Recent Changes in the National Current Employment Statistics Survey." *Monthly Labor Review* 126 (June), pp. 3–13.

Nathan, Felicia (1987). "Analyzing Employers' Costs for Wages, Salaries, and Benefits." *Monthly Labor Review* 110 (October), pp. 3–11.

National Industrial Conference Board [NICB] (1930). *Wages in the United States, 1914–1929.* New York.

——— (1939). *Management Record* (February, March).

——— (1950). *The Economic Almanac for 1950.* New York.

Nelson, Daniel (1969). *Unemployment Insurance: The American Experience, 1915–1935.* Madison: University of Wisconsin Press.

North, Simon Newton Dexter (1899). "Manufactures in the Federal Census." *Publications of the American Economic Association*, New Series, No. 2, The Federal Census. Critical Essays by Members of the American Economic Association (March), pp. 257–302.

Northrup, Herbert R., and Theresa Diss Greis (1983). "The Decline in Average Annual Hours Worked in the United States, 1947–1979." *Journal of Labor Research* 4 (Spring), pp. 95–113.

Oaxaca, Ronald L. (2000). Review of Robert A. Margo, *Wages and Labor Markets in the United States, 1820–1860. Journal of Economic History* 60 (December), pp. 1155–1156.

Officer, Lawrence H. (2007a). "An Improved Long-Run Consumer Price Index for the United States." *Historical Methods* 40 (Summer), pp. 135–147.

——— (2007b). "Value of the Consumer Bundle: A Data-Series Set." *Journal of Economic Studies* 34 (No. 3), pp. 160–178.

——— (2008a). *The Annual Consumer Price Index for the United States, 1774–2007.* MeasuringWorth. URL: http://www.measuringworth.org/uscpi.

——— (2008b). *What Was the Value of the US Consumer Bundle Then?* MeasuringWorth. URL: http://www.measuringworth.org/consumer.

Officer, Lawrence H., and Samuel H. Williamson (2006). "Better Measurements of Worth." *Challenge: The Magazine of Economic Affairs* 49 (July/August), pp. 86–110.

Persons, Warren M. (1931). *Forecasting Business Cycles.* New York: Wiley.

Phelps Brown, Ernest H., and Sheila V. Hopkins (1950). "The Course of Wage-Rates in Five Countries, 1860–1939." *Oxford Economic Papers* 2 (June), pp. 226–296.

Ransom, Roger L., Richard Sutch, and Samuel H. Williamson [RSW] (1993). "Inventing Pensions: The Origins of the Company-Provided Pension in the United States, 1900–1940." In K. Warner Schaie and W. Andrew Achenbaum, eds., *Societal Impact on Aging: Historical Perspectives*, pp. 1–38. New York: Springer.

Ravn, Morten O., and Harold Uhlig (2002). "On Adjusting the Hodrick-Prescott Filter for the Frequency of Observations." *Review of Economics and Statistics* 84 (May), pp. 371–376.

Reede, Arthur H. (1947). *Adequacy of Workmen's Compensation.* Cambridge, MA: Harvard University Press.

Rees, Albert (1959). "Patterns of Wages, Prices and Productivity." In American Assembly, *Wages, Prices, Profits, and Productivity*, pp. 11–35. New York: Columbia University.

——— (1960). *New Measures of Wage-Earner Compensation in Manufacturing, 1914–57.* New York: National Bureau of Economic Research.

——— (1961). *Real Wages in Manufacturing, 1890–1914.* National Bureau of Economic Research. Princeton: Princeton University Press.

——— (1979). "Douglas on Wages and the Supply of Labor." *Journal of Political Economy* 87 (October), pp. 915–922.

Rhode, Paul W., and Richard Sutch (2006). "National Product before 1929." In Carter et al. (2006), pp. 3.57–69.

Romer, Christina D. (1986). "Is the Stabilization of the Postwar Economy a Figment of the Data?" *American Economic Review* 76 (June), pp. 314–334.

Rosenbloom, Joshua L. (1990). "One Market or Many? Labor Market Integration in the Late Nineteenth-Century United States." *Journal of Economic History* 50 (March), pp. 85–107.

——— (1998). "The Extent of the Labor Market in the United States, 1870–1914." *Social Science History* 22 (Autumn), pp. 287–318.

——— (2000). Review of Robert A. Margo, *Wages and Labor Markets in the United States, 1820–1860*. EH.Net. URL: http://eh.net/bookreviews/library/0266.

Rubinow, Isaac M. (1914). "The Recent Trend of Real Wages." *American Economic Review* 4 (December), pp. 793–817.

Ruser, John W. (2001). "The Employment Cost Index: What Is It?" *Monthly Labor Review* 124 (September), pp. 3–16.

Samuels, Norman J. (1971). "New Hourly Earnings Index." *Monthly Labor Review* 94 (December), pp. 66–67.

Schwenk, Albert E. (1985). "Introducing New Weights for the Employment Cost Index." *Monthly Labor Review* 108 (June), pp. 22–27.

——— (1990). "Employment Cost Index Rebased to June 1989." *Monthly Labor Review* 113 (April), pp. 38–39.

Secretary of the Interior (1865). *Manufactures of the United States in 1860.* Eighth Census. Washington, DC: Government Printing Office.

——— (1866). *Statistics of the United States in 1860.* Eighth Census. Washington, DC: Government Printing Office.

Sheifer, Victor J. (1975). "Employment Cost Index: A Measure of Change in the Price of Labor." *Monthly Labor Review* 98 (July), pp. 3–12.

Shiells, Martha Ellen Koopman (1985). *Hours of Work and Shiftwork in the Early Industrial Labor Markets of Great Britain, the United States and Japan.* Ph.D. dissertation. Ann Arbor: University of Michigan.

Smith, Walter B. (1963). "Wage Rates on the Erie Canal, 1828–1881." *Journal of Economic History* 23 (September), pp. 298–311.

Sokoloff, Kenneth L. (1982). *Industrialization and the Growth of the Manufacturing Sector in the Northeast, 1820–1850.* Ph.D. dissertation. Cambridge, MA: Harvard University.

——— (1986). "Productivity Growth in Manufacturing during Early Industrialization: Evidence from the American Northeast, 1820–1860."

In Stanley L. Engerman and Robert E. Gallman, eds., *Long-Term Factors in American Economic Growth,* pp. 679–736. Chicago: University of Chicago Press.

Sokoloff, Kenneth L., and Georgia C. Villaflor [SV] (1992). "The Market for Manufacturing Workers during Early Industrialization: the American Northeast, 1820 to 1860." In Claudia Goldin and Hugh Rockoff, eds., *Strategic Factors in Nineteenth Century American Economic History,* pp. 29–65. Chicago: University of Chicago Press.

Sundstrom, William A. (2006a). "Average Daily Hours Worked and Annual Operating Days, by Manufacturing Industry: 1890–1914." In Carter et al. (2006), pp. 2.302–303.

——— (2006b). "Average Daily Hours Worked in Manufacturing, and the Distribution of Manufacturing Establishments, by Hours Worked: 1830–1890." In Carter et al. (2006), p. 2.301.

——— (2006c). "Average Weekly Hours of Production or Nonsupervisory Workers on Private Nonagricultural Payrolls, by Industry: 1909–1997." In Carter et al. (2006), pp. 2.305–307.

——— (2006d). "Average Weekly Hours of Production Workers in Manufacturing, by Sex and Degree of Skill: 1914–1948." In Carter et al. (2006), p. 2.309.

——— (2006e). "Average Weekly Hours Worked in Manufacturing, Railroads, and Bituminous Coal Mining: 1900–1957." In Carter et al. (2006), p. 2.308.

——— (2006f). "Hours and Working Conditions." In Carter et al. (2006), pp. 2.46–54.

——— (2006g). "Ratio of Hours at Work to Hours Paid: 1959–1998." In Carter et al. (2006), pp. 2.316–317.

Sutch, Richard (2006a). "Employees on Nonagricultural Payrolls, by Industry: 1900–1940 [Lebergott]. In Carter et al. (2006), pp. 2.111–112.

——— (2006b). "Employees on Nonagricultural Payrolls, by Industry: 1919–1999 [Bureau of Labor Statistics]." In Carter et al. (2006), pp. 2.112–114.

——— (2006c). "Gross Domestic Product: 1790–2002 [Continuous Annual Series]." In Carter et al. (2006), pp. 3.23–28.

U.S. Census Office (1902a). *Manufactures,* Part I, United States by Industries. Twelfth Census, Census Reports, vol. 7. Washington, DC: Government Printing Office.

——— (1902b). *Population,* Part 2. Census Reports, vol. 2. Twelfth Census. Washington, DC: Government Printing Office.

U.S. Census Bureau (2002). *Statistics for Industry Groups and Industries: 2000.* Annual Survey of Manufactures. Washington, DC: Government Printing Office.

——— (2006). *Statistics for Industry Groups and Industries: 2005.* Annual Survey of Manufactures. Washington, DC: Government Printing Office.

——— (2007). *2006 Annual Survey of Manufactures.* URL: http://factfinder.census.gov/servlet.

U.S. Chamber of Commerce (various years). *Employee Benefits*. Washington, DC: U.S. Chamber of Commerce.
——— (various years). *Employee Benefits Study*. Washington, DC: U.S. Chamber of Commerce.
——— (various years). *Fringe Benefits*. Washington, DC: U.S. Chamber of Commerce.
U.S. Department of Commerce (1954). *National Income: 1954 Edition*. Washington, DC: Government Printing Office.
Vangiezen, Robert, and Albert E. Schwenk (2001). "Compensation from before World War I through the Great Depression." *Compensation and Working Conditions* (Fall), pp. 17–22.
Walker, Francis A. (1872). *The Statistics of the Wealth and Industry of the United States*. Ninth Census, vol. 3. Washington, DC: Government Printing Office.
Weeks, Joseph D. (1886). *Report on the Statistics of Wages in Manufacturing Industries*. Census Office. Washington, DC: Government Printing Office.
Weinstein, Harriet G., and Mark A. Loewenstein (2004). "Comparing Current and Former Industry and Occupation ECEC Series." *Compensation and Working Conditions Online*.
Weiss, Thomas (1992). "U.S. Labor Force Estimates and Economic Growth, 1800–1860." In Robert E. Gallman and John Joseph Wallis, eds., *American Economic Growth and Standards of Living before the Civil War*, pp. 19–78. Chicago: University of Chicago Press.
Whaples, Robert (1990). *The Shortening of the American Work Week: An Economic and Historical Analysis of its Context, Causes, and Consequences*. Ph.D. dissertation. Philadelphia: University of Pennsylvania.
——— (2001a). "Hours of Work in U.S. History." In Robert Whaples, ed., *EH.Net Encyclopedia*, URL: http://eh.net/encyclopedia/article/whaples.work.hours.us.
——— (2001b). Review of Robert A. Margo, *Wages and Labor Markets in the United States, 1820–1860*. *Southern Economic Journal* 68 (July), pp. 200–201.
Wiatrowski, William J. (1990). "Family-Related Benefits in the Workplace." *Monthly Labor Review* 113 (March), pp. 28–33.
——— (1999). "Tracking Changes in Benefit Costs." *Compensation and Working Conditions* (Spring), pp. 32–37.
Williamson, Jeffrey G. (1975). *The Relative Costs of American Men, Skills, and Machines: A Long View*. Institute for Research on Poverty Discussion Papers. Madison: University of Wisconsin.
Williamson, Samuel H. (1992). "U.S. and Canadian Pensions before 1930: A Historical Perspective." In John A. Turner and Daniel J. Beller, eds., *Trends in Pensions: 1992*, pp. 35–57. Washington, DC: Department of Labor, Pension and Welfare Benefits Administration.
——— (1997). "The Development of Industrial Pensions in the United States during the Twentieth Century." In Gerard Caprio, Jr., and Dimitri

Vittas, eds., *Reforming Financial Systems: Historical Implications for Policy*, pp. 180–194. Cambridge: Cambridge University Press.

Wolman, Leo (1932). "American Wages." *Quarterly Journal of Economics* 46 (February), pp. 398–406.

——— (1938). *Hours of Work in American Industry*. Bulletin 71. New York: National Bureau of Economic Research.

Wood, G. Donald (1982). "Estimation Procedures for the Employment Cost Index." *Monthly Labor Review* 105 (May), pp. 40–42.

——— (1988). "Employment Cost Index Series to Replace Hourly Earnings Index." *Monthly Labor Review* 111 (July), pp. 32–35.

Wright, Carroll D. (1885). *History of Wages and Prices in Massachusetts, 1752–1883*. Boston: Wright & Potter.

——— (1900). *The History and Growth of the United States Census*. Washington, DC: Government Printing Office.

Young, Edward (1871). *Special Report on Immigration*. Executive. Document No. 1, 42nd Cong. 1st Sess. Washington, DC: Government Printing Office.

Zabler, Jeffrey F. (1972). "Further Evidence on American Wage Differentials, 1800–1830." *Explorations in Economic History* 10 (Fall), pp. 109–117.

——— (1973). "More on Wage Rates in the Iron Industry: A Reply." *Explorations in Economic History* 11 (Fall), pp. 95–99.

Index of Names

Abbott, Edith, 41–2, 43, 44, 46, 83
Abraham, Katharine G., 18, 19
Achinstein, Ascher, 78, 79–80
Adams, Donald R., Jr., 2, 10, 49, 50, 100
  real wage, 176
  wage, monthly, 144, 145, 146, 184
  wage ratio, 147
Aldrich, Nelson W., 45
Allen, Steven G., 87, 93
Atack, Jeremy, 2, 36–7, 64, 67, 99–100, 127, 136, 146
  days of operation, 52, 122, 122t, 123, 127
  hours, daily, 63, 65, 66, 133

Barkume, Anthony J., 19
Barnett, George E., 52
Bateman, Fred, 2, 37, 64, 67, 99, 100, 127, 136, 146
  days of operation, 52, 122, 122t, 123, 127
  hours, daily, 63, 65, 66, 133
Beney, M. Ada, 51, 67
  hours, weekly, 105, 106t
Bowden, Witt, 78, 79, 80, 81, 89t
  earnings, hourly, 16t, 20, 51
  earnings, weekly, 105
  hours, weekly, 55, 56, 57, 106t
Bowley, Arthur L., 83
Brandes, Stuart D., 159
Brissenden, Paul F., 35, 40, 41–2, 43, 49, 61, 62
  earnings, hourly, 76t, 77–80
  hours, weekly, 63, 106t, 132
Bullock, Charles J., 34

Calomiris, Charles W., 125
Caroll, Richard E., 25
Carter, Susan B., 1, 157, 184
Chandler, Alfred D., Jr., 47
Coelho, Philip R. P., 2, 41, 43, 44, 45, 100, 140, 151–2, 176
  real wage, 177
  wage, daily, 136, 138, 149
  wage ratio, 137, 141–3t
Coombs, Whitney, 41–2, 46, 51
Creamer, Daniel, 25–6
  wage, hourly, 76t, 81; used in Rees, 91, 93, 104

David, Paul A., 48, 51
  consumer price index, 178
  real wage, 177, 178
  wage, 1, 178
Davis, Joseph H., 2
  industrial production, 1, 124–5
Davis, Milton B., 107t
DeBow, James Dunwoody Brownson, 45, 185t
Devens, Richard M., Jr., 54
Dewey, Davis R., 41–3
Douglas, Paul H., 2, 26, 27, 28, 34, 35, 37, 49, 61, 75, 78, 79, 87, 88t, 89t, 94, 100, 123
  earnings, annual, 92t, 93, 109, 110t, 111t, 112
  earnings, hourly, 16t, 76t, 77, 80–1, 133–4, 134t, 143
  hours, weekly, 106t, 132
Douty, Harry M., 23, 26, 27, 28, 31, 46, 49, 83

INDEX OF NAMES

Easterlin, Richard A., 3, 4, 5, 39, 40, 112, 113, 124

Fabricant, Solomon, 39, 107t, 109, 111t
Falkner, Ronald P., 46
  hours, daily, 65–6, 187
  wage, daily, 81, 82t, 83–4, 134, 136; revised, 135–6
Famulari, Melissa, 71–2
Fishback, Price V., 101
  workers'-compensation history, 158, 160
Frickey, Edwin, 124–5

Gallman, Robert E., 37
  value-added, 124, 125
Getz, Patricia M., 21
Girard, Stephen, 50
Goldin, Claudia, 2, 47, 99, 100
  employment by age-sex, Northeast, 130, 188–9t
  real wage, 178, 179
  wage, adult-female/adult-male, 148t
  wage, boy/adult-male, 129, 147–8, 149
  wage, female/adult-male, 129, 147–8, 148t
Greis, Theresa Diss, 11, 53, 54, 73
Grosse, Scott D., 145

Haines, Michael R., 129
Hanes, Christopher, 1, 8, 29, 46, 49, 87, 125
  earnings, hourly, 88–9t, 94–5
Hansen, Alvin H., 77, 89t
Hopkins, Sheila V., 83, 84, 89t
  wage, daily, 82t, 84–5, 134

Jablonski, Mary, 11, 57, 59
  work-hours/paid-hours ratio, 58t, 60–1
Jacobs, Eva E,, 6
James, John A., 1, 157
Johnston, Louis D., 1

Jones, Ethel B., 40, 53, 54, 55, 67, 68, 73, 87, 156
  hours, weekly, 56, 105, 106t, 108, 132–3

Kantor, Shawn Everett, 101
  workers'-compensation history, 158, 160
Kendrick, John W., 37–8, 55, 61, 62, 111t, 112
  hours, weekly, 63, 105, 106t, 132
Kennedy, Joseph C. G., 130t
King, Willford Isbell, 52, 77
  earnings, hourly, 20–1, 105
  hours, weekly, 56–7, 68, 105, 106t, 108
Kirkland, Katie, 53–4
Kunze, Kent, 11, 57, 59
  work-hours/paid-hours ratio, 58t, 60–1

Lamberson, Frances, 75, 78
  earnings, hourly, 76t, 77
Latimer, Murray W., 80
Layer, Robert G., 49
  earnings, daily, 144
  hours, daily, 68
Lebergott, Stanley, 44–5, 46, 47, 48, 83, 84, 85–6, 87, 92, 93, 140, 190, 192
  employment, 161, 184
  wage, daily, 113, 120–1, 120t
Lettau, Michael K., 19
Lindert, Peter H., 1, 48, 145, 177
Long, Clarence D., 2, 28, 29, 30, 31, 34, 36, 39, 42–3, 44, 45, 46, 61, 65, 66, 83, 84, 85, 88t, 89t, 94, 99, 100
  days of operation, 122t, 123–4, 126–7
  earnings, annual, 111t, 112–13
  hours, daily, Aldrich, 187–8, 191–2t
  hours, daily, Weeks, 64, 191–2t
  wage, daily, Aldrich, 82t, 85–6, 136; extended, 135–6; revised, 134–5

wage, daily, *Bulletin 18*, 82t, 86, 134, 138–40
wage, daily, Weeks, 82t, 85–6, 134

McLane, Louis, 47
Magdoff, Harry, 107t
Manser, Marilyn E., 71–2
Margo, Robert A., 1, 2, 6, 9, 43, 44, 45, 46, 48, 49, 50, 50–1, 66, 76t, 80, 87, 89t, 92t, 94, 99, 100, 111t, 120, 127, 132, 139, 144, 145, 146
   days of operation, 52, 122, 122t, 123, 127
   hours, daily, 63, 133
   real wage, 176–80
   wage, daily, 149–51, 152
   wage ratio, 131, 141–3t
Millis, Harry A., 159
Miron, Jeffrey A., 124–5
Mitchell, Wesley C., 83
   wage, daily, 82t, 84, 134
Moehrle, Thomas G., 19, 73
Montgomery, Royal E., 159
Morisi, Theresa L., 19

Nathan, Felicia, 69, 72
Nelson, Daniel, 159
North, Simon Newton Dexter, 3, 112
Northrup, Herbert R., 11

Oaxaca, Ronald L., 150
Officer, Lawrence H.
   consumer price index, 1, 169, 171
   value of consumer bundle, 172
Otto, Phyllis Flohr, 11, 57, 59
   work-hours/paid-hours ratio, 58t, 60–1

Persons, Warren M., 124–5
Phelps Brown, Ernest H., 83, 84, 89t
   wage, daily, 82t, 84–5, 134

Ransom, Roger L., 159
Ravn, Morton O., 125, 126

Reede, Arthur H., 101
   workers' compensation, 160
Rees, Albert, 2, 7, 10, 11, 19, 20–1, 26, 27, 28, 36, 38, 39, 40, 42, 43, 49, 51, 52, 53, 54, 55, 56–7, 68, 71, 73, 75, 76t, 77, 78, 79, 80, 81, 89t, 98–9, 100, 101, 105
   benefits, 156, 157, 163
   days of operation, 90, 121, 122–3, 122t
   earnings, annual, 87, 87t, 90, 92t, 93, 109, 110t, 111t, 112, 113, 134, 134t
   earnings, daily, 90
   earnings, hourly, 87, 87t, 88t, 90, 91–4, 103, 104, 133–4
   hours, daily, 90, 132–3, 174
   hours, weekly, 63, 90, 91, 105, 106t, 107t, 108
   work-hours/paid-hours ratio, 91
Rhode, Paul W., 1, 157
Romer, Christina D., 124–5
Rosenbloom, Joshua L., 28, 29, 30, 138, 150
Rubinow, Isaac M., 78
   earnings, hourly, 16t, 24t, 75, 76–7, 76t
Ruser, John W., 25

Samuels, Norman J., 25
Schwenk, Albert E., 25, 26, 27
   group insurance, 160
Sheifer, Victor J., 25
Shepherd, James F., 2, 41, 43, 44, 45, 100, 140, 151–2, 176
   real wage, 177
   wage, daily, 136, 138, 149
   wage ratio, 137, 141–3t
Shiells, Martha Ellen Koopman, 107t
Siegel, Irving H., 107t
Smith, Walter B., 50, 144
Sokoloff, Kenneth L., 2, 10, 36, 37, 47–8, 99, 100, 133, 146
   days of operation, Northeast, 127–8

Sokoloff, Kenneth L.—*Continued*
   employment by age-sex, Northeast, 130, 188–9t
   wage, adult-female/adult-male, 148t
   wage, annual, Northeast, 113, 120–1, 120t
   wage, boy/adult-male, 129, 147–8, 149
   wage, female/adult-male, 129, 147–8, 148t
Solar, Peter, 48, 51
   consumer price index, 178
   real wage, 177, 178
   wage, 1, 178
Spletzer, James R., 18, 19
Sprague, Shawn, 58t
Steuart, William M., 40
Stewart, Jay C., 18, 19
Sundstrom, William A., 53–4, 57, 58t, 59, 60–1, 65, 66, 67, 87, 94, 105, 107t
   hours, daily, Weeks, 64–5, 190, 191–2t
Sutch, Richard, 1, 157, 161, 177
   pensions history, 159
Sylla, Richard, 1, 157

Thomasson, Melissa A., 158

Uhlig, Harold, 125, 126

Vangiezen, Robert, 26, 27
   group insurance, 160

Villaflor, Georgia C., 2, 10, 50–1, 99, 139, 146
   days of operation, Northeast, 127–8
   wage, annual, Northeast, 113, 120–1, 120t
   wage, daily, 149–50, 151

Walker, Francis A., 116t, 119t, 182
Weeks, Joseph D., 43–5, 63–5, 82t, 100
Weiss, Thomas, 129–30
Whaples, Robert, 65, 66, 67, 105, 107t, 108, 150–1
   hours, daily, 133, 191–2t, 192
Wiatrowski, William J., 23, 69, 71, 72, 163
Williamson, Jeffrey G., 48, 85, 86, 145
Williamson, Samuel H., 1, 2, 101
   pensions history, 159, 161
   value of consumer bundle, 172
Wolman, Leo, 10, 63, 67–8, 80, 81, 89t
   hours, weekly, 106t
Wood, G. Donald, 25
Wright, Caroll D., 36, 37, 46, 48, 52, 190
   wage, daily, 144

Young, Edward, 47

Zabler, Jeffrey F., 2, 49–50, 100
   wage, monthly, 144–6, 184

Index of Subjects

agricultural implements, 134–5, 182, 184, 190
agriculture, 3, 145
 see also agriculture processing; agriculture services
agriculture processing, 39, 111t, 112, 117t, 185t
agriculture services, 116t
Aldrich Report, 45–7, 65–7, 81–3, 84, 85, 187, 191–2t
ale, beer, porter, 134–5, 182, 190
American Factfinder Help, 3
annual earnings,
 see earnings, annual
Annual Survey of Manufactures, 31–3, 98
 see also Census Bureau
annual wage, see wage, annual
Army civilian employees, 50–1, 144, 149–52, 176–7
artisans, see skilled workers
average annual earnings,
 see earnings, annual
average annual wage,
 see wage, annual
average daily earnings,
 see earnings, daily
average daily hours,
 see hours, daily
average daily wage, see wage, daily
average hourly benefits,
 see benefits, hourly
average hourly earnings,
 see earnings, hourly
average hourly wage,
 see wage, hourly

average monthly hours,
 see hours, monthly
average monthly wage,
 see wage, monthly
average number of days of operation,
 see days of operation
average weekly earnings,
 see earnings, weekly
average weekly hours,
 see hours, weekly

bakeries, 33, 114t, 116t, 130t, 134–5
Baltimore, 30
banksman, 145
bellhanging, 114t, 116t, 118t, 185t
benefits
 Bureau of Economic Analysis, 70t, 73–4, 156, 157t
 Census Bureau, 70t, 74
 Chamber of Commerce, see U.S. Chamber of Commerce
 components, 9
 Department of Commerce, 157–8
 employer cost versus employee valuation, 71–2
 Employer Costs for Employee Compensation, 69–71, 72, 100–1, 155–6
 Employer Expenditures for Employee Compensation, 70t, 72, 156
 Employment Cost Index, 69
 evolution over time, 1–2, 9, 162–3, 165–9

benefits—*Continued*
　markup over compensation, 165–8
　markup over earnings, 12, 71, 98, 100, 155, 156–7, 165
　pension-and-welfare, 101, 158–60, 161–2
　Rees, 101, 156
　relationship to earnings concept, 12–13, 155
　U.S. Chamber of Commerce, 70t, 72–3
　workers' compensation, 101, 157–8, 160–1, 161–2
　see also benefits, hourly; group insurance; health and welfare programs; paid leave; pensions; profit-sharing; stock ownership; supplemental pay; unemployment insurance
benefits, hourly, 9
　1900–2006, 165–9
　1900–1928, 101, 157–63
　1929–2006, 100–1, 155–7
　see also benefits
bicycle and tricycle repairing, 114t, 117t
blacksmith, 27, 47, 114t, 116t, 118t, 130t, 136, 139, 145, 182t, 185t
bleaching, dyeing, cleaning, 114t, 116t, 118t, 130t, 185t
BLS, *see* Bureau of Labor Statistics
blue-collar workers,
　see production workers
boiler maker, 139, 182t
books and newspapers, 183, 184, 190
boots and shoes, 111t, 112, 115t
Boston, 30
bottling, 114t, 118t
boys, 128–9, 149, 186–7, 188t
Brandywine Region, 94, 144
bridge construction,
　see construction

building trades, 27, 76, 82t, 83, 84
　see also carpenter; construction
Bulletin 18, 28–31, 85, 136, 138–9, 140–1
Bulletin 77, 16t, 26–7, 61
Bulletin 93, 40–1, 78, 79
Bureau of Economic Analysis, 70t, 73–4, 101, 156, 157t
Bureau of Labor, 26, 28
　see also Commissioner of Labor
Bureau of Labor Statistics, 15
　production-workers definition, 5–6
　production-workers termination, 21–2
　see also Bulletin 18; *Bulletin 77*; Bureau of Labor; Commissioner of Labor; Composition of Payroll Hours in Manufacturing; Current Employment Statistics Survey; Employer Costs for Employee Compensation; Employer Expenditures for Employee Compensation; Employment Cost Index; Hourly Earnings Index; Hours-at-Work Survey; *Monthly Labor Review*; payroll industries; union industries
Bureau of the Census, *see* Census Bureau
butchering, 114t, 118t

cabinet maker, 115t, 139, 182t
California, 150
　see also San Francisco
Canada, 161
canal maintenance, 50, 144
candy stores, 33
carpenter, 45, 47, 116t, 118t, 136, 145
　ship, 84
carriages and wagons, 134–5, 183–4

INDEX OF SUBJECTS ❖ 215

cars, railroad, omnibuses, and repairing, 39, 112–13, 117t, 118t
carving, 114t, 116t, 185t
cement, 116t, 118t
Census, *see* Census Bureau
Census Bureau
   benefits, 70t, 74
   factory-system definition, 3–4
   manufacturing definition, 3
   preferred data source, 12–13, 98
   production-workers continuation, 22
   production-workers definition, 5
   see also American Factfinder Help; Annual Survey of Manufactures; *Bulletin 93*; Census of Manufactures; Census of Social Statistics; Dewey Report; Secretary of the Interior; Weeks Report
Census of Manufactures
   days of operation, 52, 121–4
   earnings, annual, 33–40, 76t, 78, 87–90, 99, 109–19, 130
   earnings, weekly, 20, 54–5
   employment, 2, 131, 135, 136, 152, 182–6
   hours, daily, 63, 133
   hours, weekly, 62–3, 90, 91, 105, 106t, 108
   persons engaged in *Bulletin 18* occupations, 139–40, 181–2
   quality of data, 36–8
   size cutoff, 38
   years covered, 35–6
   see also Census Bureau
Census of Social Statistics, 45, 150
Census Office, *see* Census Bureau
Chamber of Commerce, *see* U.S. Chamber of Commerce
chemical manufacturing, *see* white lead
Chicago, 30
children, 188t
   see also boys; girls

Cincinnati, 30
city public works, *see* public works, city
civilian employees of U.S. Army, *see* Army civilian employees
cleaning, *see* bleaching, dyeing, cleaning
clerk, 5, 34, 82t, 83, 145
   see also nonproduction workers
clothing, 80, 111t, 112, 115t, 189t
   see also tailor
coffins, 114t, 117t, 118t
collier, 145
Commissioner of Labor, 31, 61, 106–7t
common laborer, *see* laborer
compensation, average hourly, *see* compensation, hourly
compensation, hourly, 165–9
   actual-work-hour versus paid-hour foundation, 11
   components, 9, 165
   growth, 165, 168f
   methodology, 6–11
compensation, nominal, *see* compensation, hourly
compensation, real, *see* real wage
Composition of Payroll Hours in Manufacturing (work-hours/paid-hours ratio), 58t, 60
compositor, 139, 182t
Connecticut, 89t, 92t, 129–30
construction, 3, 30, 39, 43, 109, 116t, 117t, 130t, 140, 150, 185t
   bridge, 39, 116t, 118t
   railroad, 31
   sidewalks, 82t
   see also building trades; carpenter
consumer bundle, 172–5
consumer price index, 169–72, 178
cotton, *see* cotton manufactures
cotton manufactures, 82t, 114t, 116t, 118t, 134–5, 183, 184
cotton textiles, *see* cotton manufactures

Current Employment
Statistics Survey (and
predecessors)
earnings, 1904–1931, 15, 16t,
20–1, 27–8, 104
earnings, 1932 onward,
15–19, 21–2
hours, weekly, prior to 1932,
54–7
hours, weekly, 1932 onward,
53–4, 105
see also Hourly Earnings Index
custom trades, 3–4, 38, 109,
110–11t, 112, 113, 114–19t
see also hand trades

daily earnings, *see* earnings, daily
daily hours, *see* hours, daily
daily wage, *see* wage, daily
dairy products, 115t
days of operation, 15, 52–3, 121–8
days of operation, average number,
see days of operation
Delaware, 136, 138, 186t
dentistry, 40, 116t, 118t
Department of Commerce, 157–8
Department of Commerce and
Labor, 15
Department of Labor, 15, 28
see also Bulletin 18; Bureau of
Labor; Bureau of
Labor Statistics
Dewey Report, 41–3
District of Columbia, 64, 136,
138, 186t
draperies, custom, 33
dry goods, 82t, 83
DuPont, 49, 144, 146
dyeing, *see* bleaching, dyeing,
cleaning

earnings
current-weight, 6–7, 9, 97
definition, 6
denomination, 8–9
fixed-weight, 6–7, 9

gross, 11–13, 17, 23, 32, 34, 40,
51, 52, 97
regular, 11–12, 22
time dimension, 8
see also earnings, annual;
earnings, daily; earnings,
hourly; earnings, weekly;
paid leave; supplemental pay
earnings, annual
Census of Manufactures, 33–40
Douglas, 92t, 93, 109–12, 134
Long, 111t, 112–13
new series, 99, 112–13, 114–19t,
127–32
Rees, 87–90, 92t, 93, 99,
109–12, 134
state labor bureaus, 48–9
see also wage, annual
earnings, daily
Layer, 144
new series, 99, 127–46
Rees, 88t, 90
see also wage, daily
earnings, hourly
Bowden, 76t, 77–80
Bureau of Labor Statistics,
15–25, 27–8
Census, 31–3, 40
Douglas, 76t, 80–1, 133–4, 153
Hanes, 88–9t, 94–5
King (National Bureau of
Economic Research), 52
National Industrial Conference
Board, 51, 103–4
new series, 98–100, 103–5,
108–9, 146, 147–53
Rees, 88t, 90–4, 98–9, 103, 104,
133–4
see also wage, hourly
earnings, regional
Northeast, 130
see also wage, regional
earnings, weekly
Census (*Bulletin 93*), 40–1
Dewey, 41–3
see also wage, weekly

INDEX OF SUBJECTS ♦ 217

earnings ratio, rest-of-U.S./
    Northeast, 131, 152
    see also wage ratio,
    U.S./Northeast
East North Central, 138
East South Central, 45
Economic Census—Manufacturing,
    see Census Bureau
electric light and power, 117t, 118t
Employer Costs for Employee
    Compensation
    benefits, 69–71, 72
    benefits markup, 155–6
    earnings, 12, 16t, 22–3, 103
Employer Expenditures for
    Employee Compensation
    benefits, 70t, 72
    benefits markup, 156
    earnings, 16t, 23
    work-hours/paid-hours ratio, 60
employment
    adult-females/all-workers, 188t
    adult-males/all-workers,
        186–7, 188t
    adult-males/males, 149, 187
    boys/all-workers, 186–7, 188t
    boys/males, 149, 187
    Census of Manufactures, 135,
        136, 140–1, 181–6
    children/all-workers, 188t
    females, young, 188t
    females/all-workers, 149,
        186–7, 188t
    females-and-boys/all-workers,
        188–9t
    girls/females, 148t, 188–9t
    Lebergott
        (textiles, iron and steel), 146
    males/all-workers, 149,
        186–7, 188t
    manufacturing, total, 2, 78–9
    Northeast, 152, 184–6
    rest of U.S., 152, 184–6
    union workers in New York, 78
Employment Cost Index
    benefits, 69

earnings, 16t, 24–5, 103
    work-hours/paid-hours ratio,
        57–9
engineer, 136
Erie Canal, 50, 144
establishment payroll survey,
    see Current Employment
    Statistics Survey
establishment survey, see Current
    Employment Statistics
    Survey
establishments, full-time, see full-
    time establishments
establishments, part-time, see part-
    time establishments

factories
    versus hand trades, 3–4, 38–9
    versus neighborhood
        establishments, 4
    see also manufacturing
factory establishments, see factories
females, 128–30, 147–8, 149,
    186–7, 188t
    see also females, adult; girls
females, adult, 144, 148t, 188t
filler, 145
fishing, 3, 111t, 112, 116t,
    130t, 185t
flax dressing, 112–13
Florida, 150–1
forestry, 3, 39, 111t, 112, 116t,
    130t, 185t
    see also logging; lumber
foundries, 51
full-time establishments, 127–8
furniture, 115
    see also cabinet maker

gainfully employed, see employment
gas, illuminating and heating,
    see illuminating gas
gilding, 115t
girls, 148t, 188t
glass, 109
grindstones and millstones, 112–13

groceries, 82t
group insurance, 160
gunpowder, 146
gunsmithing, 114t, 118t

hand trades, 3–4, 38–9, 109, 110–11t, 112, 113, 114–19t
health and welfare programs, 159–60
heating, *see* illuminating gas
Hodrick-Prescott filter
  description, 125–6
  real wage, 126–7
  value-added, 178–9
home production, 4, 38
horseshoeing, *see* blacksmith
hourly benefits, *see* benefits, hourly
hourly earnings, *see* earnings, hourly
Hourly Earnings Index, 16t, 24t, 25
hourly wage, *see* wage, hourly
hours
  actual-work, 10, 11, 54, 57, 63, 66–7, 68, 92–3, 97, 105
  at-work, *see* actual-work *under this main level*
  full-time, 10, 20, 61, 62, 63, 64, 65, 66–7, 68, 78, 92–3, 97
  nominal, *see* full-time
    under this main level
  normal, *see* full-time
    under this main level
  paid, 11, 53–4, 57, 63, 97
  paid-leave, 11
  prevailing, *see* full-time
    under this main level
  regular, *see* full-time
    under this main level
  scheduled, *see* full-time
    under this main level
  standard, *see* full-time
    under this main level
  straight-time, *see* full-time
    under this main level
  work, *see* actual-work
    under this main level
  work-hour, *see* actual-work
    under this main level
worktime, 10–11
hours, daily, 15, 53
  Aldrich Report, *see* Falkner
    under this main level
  Atack, Bateman, and Margo, 63, 99–100, 133
  Atack and Bateman, 63, 65, 66, 99–100, 133
  Commissioner of Labor, 61
  Falkner, 65–6, 100, 187, 191–2t
  Layer, 69
  Lebergott, 190, 193
  Long, 61, 89t, 187, 191–2t
  McLane Report, 66–7, 133
  Margo, 191–2t
  new series, 99–100, 132–3, 187, 190–3
  Rees, 88–9t, 90
  state labor bureaus, 67
  Sundstrom, 190, 191–2t
  Weeks Report, 63–5, 66, 100, 187, 191–2t
  Whaples, 191–2t
  Wright, 190
hours, monthly
  (Man-Hour Statistics), 62
hours, weekly, 10
  Beney, *see* National Industrial Conference Board
  Bowden, 106t
  Brissenden, 63, 78, 106t, 132
  Bulletin 77, 61
  Bureau of Labor Statistics
    surveys 1907–1918, 61
    see also Current Employment Statistics Survey
  Census Bureau, 62–3, 90, 108
  Current Employment Statistics Survey, 53–7, 105, 106t
  Douglas, 90, 106t, 132
  Douglas and Lamberson, 78
  Jones, 105, 106t, 108, 132–3
  Kendrick, 63, 105, 106t, 132
  King, 68, 105, 106–7t

National Bureau of Economic
  Research, *see* King
  under this main level
National Industrial Conference
  Board, 67–8, 105, 106t, 108
National Safety Council, 68
  Rees, 63, 105, 106t, 108, 132
  Rubinow, 78
  Wolman, 63, 90, 106t, 133
Hours-at-Work Survey (work-
  hours/paid-hours ratio), 58t,
  59–60, 103
hours-worked/hours-paid ratio, *see*
  work-hours/paid-hours ratio

Illinois, 94
  see also Chicago
illuminating gas, 34, 112–13, 117t,
  134–5, 183
industrial production, 124–5
Iowa, 92t, 122t
iron, *see* iron and steel
iron and steel, 80, 139, 144–5, 146,
  182, 183, 184
  see also metals and metallic goods
iron-and-steel workers,
  see iron and steel
iron molder, *see* iron and steel

jewelry repairing, *see* watch, clock,
  and jewelry repairing

Kansas, 150–1
keeper, 145
kindling wood, *see* lumber

laborer, 30, 45, 136, 139, 145,
  150, 182t
laundry work, 40, 116t
leather, 134–5, 183, 184, 190
locksmithing, 114t, 116t,
  118t, 185t
logging, 33, 39, 87, 92t, 109,
  110t, 116t
  see also forestry; lumber
Lowell (Massachusetts), 69

lumber, 82t, 83, 85, 87, 109, 114t,
  116t, 118t, 185t
  see also forestry; logging

machine shops, 51
machinist, 136, 139, 182t
McLane Report, 47–8, 66–7, 99,
  120, 133, 192–3
Maine, 89t, 122t
males, 146–7, 149, 186–7, 188t
  see also boys; males, adult
males, adult, 120–1, 128–9, 147–9,
  186–7, 188t
Management and Budget,
  Office of, 12
Man-Hour Statistics
  earnings, hourly, 40
  hours, monthly, 63, 174
manual workers, *see* production
  workers
manufactured gas,
  see illuminating gas
manufacturing
  definition, 2–4
  value-added, 124–5
  versus other sectors, 39–40
  see also factories
manufacturing production,
  see industrial production
Maryland, 122t, 136, 138, 186t
  see also Baltimore
Massachusetts, 47, 48, 53, 87–8,
  89t, 90, 92t, 122t, 123,
  129–30, 144, 148t
  see also Boston;
    Lowell (Massachusetts)
medical benefits, 159
men's clothing, *see* clothing
metals and metallic goods, 134–5,
  183, 184
  see also iron and steel
Michigan, 94
Middle Atlantic, 120t, 129–30,
  136, 137, 138, 185–6t
  see also individual states
Middle States, *see* Middle Atlantic

Midwest, 138, 150–1, 152
    see also individual states
miller, 145
millinery, see clothing
mining, 3, 39, 43, 111t, 112, 116t, 130t, 140, 185t
    iron, 84
monthly hours, see hours, monthly
Monthly Labor Review (wage-rate changes), 25–6
monthly wage, see wage, monthly
Mountain Region, 45, 138

NAICS, see North American Industry Classification System
National Bureau of Economic Research (King), 20–1, 52, 56–7, 68, 77, 105, 106t, 108
National Compensation Survey, see Employer Costs for Employee Compensation; Employer Expenditures for Employee Compensation; Employment Cost Index
National Industrial Conference Board, 51, 67–8, 88–9t, 91, 95, 105, 106–7t, 108
National Safety Council, 68
neighborhood establishments, see custom trades
neighborhood trades, see custom trades
New England, 47, 120t, 129–30, 137, 138, 148t, 185–6t
    see also individual states
New Hampshire, 129–30
New Jersey, 53, 87–8, 89t, 90, 92t, 122t, 123, 129–30
New Orleans, 30
New York (city), 30
New York (state), 47, 49, 78, 79, 92t, 94, 129–30
    see also New York (city)
newspapers, see books and newspapers

nominal compensation, see compensation, hourly
nonmonetary earnings, see benefits
nonmonetary wages, see benefits
nonoffice workers, see production workers
nonproduction workers, 145, 150
    definition, 4–5
North American Industry Classification System, 16t, 17, 23, 33, 70t, 155–6
Northeast, 44–5, 47, 97–8, 120t, 128, 130–2, 136, 138, 140, 146, 147–9, 150, 151–2, 176, 181–2, 184, 185, 186, 188–9t
    see also Brandywine Region; Middle Atlantic; New England

Office of Management and Budget, see Management and Budget, Office of
office workers, see nonproduction workers
Ohio, 47, 89t, 92t, 122t, 136
    see also Cincinnati
onsite medicine, 159

Pacific Region, 45, 138
    see also California
paid leave, 12, 155
paid-hours/work-hours ratio, see work-hours/paid-hours ratio
painter, 116t, 119t, 136
paper, 134–5, 183, 184, 190
part-time establishments, 127–8, 132
part-time workers, 40, 127–8
pattern maker, iron works, see iron and steel
paving, 118t
payroll industries, 27–8, 76, 77, 80, 81, 100, 133–4

INDEX OF SUBJECTS   221

Pennsylvania, 47, 49–50, 53, 87–8, 89t, 90, 92t, 122t, 123, 129–30, 144–5
  see also Philadelphia; Pittsburgh
pension-and-welfare benefits,
  see benefits
pensions, 101, 159
Philadelphia, 30, 144, 176
photography, 114t, 116t, 118t, 185t
piecework earnings, 5, 27
Pittsburgh, 30
plumbing, 116t, 118t
printing, see books and newspapers
production workers
  Bureau of Labor Statistics termination of, 21–2
  Census continuation of, 22
  combined with nonproduction workers, 69, 70t, 155–6
  definition, 4–6
  importance, 2
  profit-sharing, 159–60
  public works, city, 82t, 83
  see also street and sewer work
publishing, 33
  see also books and newspapers

quarrying, 116t, 130t, 185t

railroad cars, see cars, railroad, omnibuses, and repairing
railroad construction,
  see construction
railroad repair shops, see cars, railroad, omnibuses, and repairing
railroads, 82t, 83, 84
  see also cars, railroad, omnibuses, and repairing; construction
real compensation, see real wage
real hourly compensation,
  see real wage
real wage, 169–72
real wage, comparison with
  Adams, 176
  Coelho and Shepherd, 177

David and Solar (per Margo), 177–9
Goldin, 178, 179
Margo, 176–7, 179–80
Regular Census, see Census of Manufactures
Reports of Persons and Articles Hired, see Army civilian employees
Rhode Island, 89t, 129–30
Richmond, 30
rigging, 114t, 116t, 130t, 185t
roofing, 116t, 118t

saddler, 47
St. Louis, 30
St. Paul, 30
salaried workers,
  see nonproduction workers
salaries, 5
San Francisco, 30
Secretary of the Interior, 186t
semi-skilled workers,
  see unskilled workers
services, 3, 116t, 117t, 185t
  see also dentistry; laundry work
sewing-machine repairing, 115t
shoes, see boots and shoes
SIC system, see Standard Industrial Classification System
sidewalks construction,
  see construction
skilled workers, 13, 85, 86, 140, 144, 145, 150
smith, see blacksmith
Social Statistics,
  see Census of Social Statistics
South, 30–1, 45, 51, 67–8, 105, 138
  see also East South Central; South Atlantic; West South Central; individual states
South Atlantic, 150–1, 152
  see also individual states
South Carolina, 89t, 92t
South Central, 150, 151, 152

Standard Industrial Classification System, 16t, 17, 23, 33, 70t, 155–6
standard of living, *see* consumer bundle; real wage; real wage, comparison with; work-hours
state labor bureaus
  days of operation, 52–3, 88–9t, 90, 121–4
  earnings, annual, 48–9, 87, 88–9t, 90, 92t, 93
  unemployment (New York), 78, 79
  wage rates,
    daily (Massachusetts), 48
stationery, 117t
steel, *see* iron and steel
stock ownership, 159–60
stone, 31, 82t, 83, 84, 85
stonecutter, 139, 182t
street and sewer work, 27
  see also public works, city
supplemental pay, 12, 155
supplements to wages and salaries, *see* benefits

tailor, 33
  see also clothing
taxidermy, 114t, 118t
teamster, 30, 139, 145, 150, 182t
textiles, 46, 124, 146, 184
  see also cotton manufactures; woolen manufactures
timber cutting, *see* logging; lumber; woodcutter
tinplate and terneplate, 109
tire retreading, 33
tobacco, 114t, 118t
trade (wholesale and retail), 3, 40

unemployment insurance, 159
union industries, 27–8, 76, 77, 80–1, 100, 133
United States Census Office, *see* Census Bureau

unskilled workers, 13, 140, 144, 145, 150
upholstery, 114t, 115t, 116t, 118t, 186t
U.S. Army, civilian employees, *see* Army civilian employees
U.S. Census Bureau, *see* Census Bureau
U.S. Chamber of Commerce, 72–3

wage, annual
  Sokoloff and Villaflor, 99, 113, 120–1, 127–8
  see also earnings, annual
wage, daily
  Adams, 50
  Aldrich Report, 45–7
  Bureau of Labor (*Bulletin 18*), 28–31, 140–3
  Census (Social Statistics), 45
  Coelho and Shepherd, 136–8, 141–3t
  Falkner, 81–4, 135
  Lebergott, 113, 120–1
  Long, 82t, 85–6, 134–5, 138–40, 153
  McLane Report, 47–8
  Margo, 50–1, 141–3t, 144, 149–52
  Margo and Villaflor, 50–1
  Mitchell, 82t, 84
  Phelps Brown and Hopkins, 82t, 84–5
  Smith, 50, 144
  Weeks Report, 43–5
  Wright, 48, 144
  see also earnings, daily
wage, hourly
  Bureau of Labor, 16t, 24t, 26–7
  Bureau of Labor Statistics, 25–6, 27–8
  Commissioner of Labor, *see* Bureau of Labor *under this main level*
  Creamer, 76t, 81
  Department of Labor, *see* Bureau of Labor *under this main level*

Dewey, 41–3
Douglas and Lamberson, 76t, 77
Hanes, 88t, 94
Rubinow, 75–7
   see also earnings, hourly
wage, monthly
   Adams, 49, 144, 145, 146, 147
   Zabler, 49–50, 144–6
wage, regional
   California, 150
   Midwest, 138, 150–1
   Mountain, 138
   Northeast, 113, 120–1, 130, 136, 138, 140, 151–2
   Pacific, 138
   rest of U.S. (after Northeast), 138, 140, 151–2
   South, 138
   South Atlantic, 150–1
   South Central, 151
   West, 138
   see also earnings, regional
wage, weekly
   Young Report, 47
   see also earnings, weekly
wage rate, 4–5
   definition, 6
   see also wage, annual; wage, daily; wage, hourly; wage, monthly
wage ratio, Northeast
   adult-female/adult-male, 148t
   all-worker/male, 141t, 149
   boy/adult-male, 128–9, 149
   female/adult-female, 148t
   female/adult-male, 128–9, 147–8
   female/male, 147, 149
   girl/adult-female, 148t
   male/adult-male, 149
wage ratio, skilled/unskilled, 145
wage ratio, U.S./Northeast, 131–2, 136–43
   see also earnings ratio, rest-of-U.S./Northeast
wage supplements, see benefits
wage-earners,
   see production workers

wage-rate index, average hourly,
   see wage, hourly
wagons, see carriages and wagons
watch, clock, and jewelry repairing, 114t, 116t, 118t, 186t
weaving, 115t
weekly earnings,
   see earnings, weekly
weekly hours, see hours, weekly
Weeks Report, 43–5, 46–7, 63–5, 66, 84, 85, 136, 190, 191–2t
welfare programs, see health and welfare programs
West, 30–1, 138
   see also Mountain Region; Pacific Region
West South Central, 138
wheelwrighting, 114t, 116t, 118t, 130t
white lead, 83, 134–5, 183, 184, 190
white-collar workers,
   see nonproduction workers
whitesmithing, 114t, 116t
Wisconsin, 89t, 92t
woodcutter, 114t, 116t, 145
   see also logging; lumber
woodcutting and cording,
   see lumber; woodcutter
woolen manufactures, 134–5, 183, 184, 190
workers, nonproduction,
   see nonproduction workers
workers, production,
   see production workers
workers, semi-skilled,
   see unskilled workers
workers, skilled, see skilled workers
workers, unskilled,
   see unskilled workers
workers' compensation, see benefits
work-hours
   actual, 174–5
   full-time workyear, 172, 174
   required to purchase consumer bundle, 172–5

work-hours/paid-hours ratio,
 57–60
 Composition of Payroll
  Hours in Manufacturing,
  58t, 60
 Employer Expenditures for
  Employee Compensation,
  58t, 60
 Employment Cost Index, 57–9

Hours-at-Work Survey, 58t,
 59–60, 103
Jablonski, Kunze, and Otto,
 58t, 60–1
National Compensation Survey,
 see Employment Cost Index
Rees, 88–9t, 91

Young Report, 47